Early Mathematics Learning and Development

Series editor

Lyn D. English
Queensland University of Technology, School of STM Education
Brisbane, QLD, Australia

More information about this series at http://www.springer.com/series/11651

Virginia Kinnear · Mun Yee Lai
Tracey Muir
Editors

Forging Connections in Early Mathematics Teaching and Learning

Springer

Editors
Virginia Kinnear
School of Education, College of Education, Psychology and Social Work
Flinders University
Bedford Park, SA
Australia

Tracey Muir
Faculty of Education, College of Arts, Law and Education
University of Tasmania
Launceston, TAS
Australia

Mun Yee Lai
Faculty of Education and Arts, School of Education
Australian Catholic University
Fitzroy, VIC
Australia

ISSN 2213-9273 ISSN 2213-9281 (electronic)
Early Mathematics Learning and Development
ISBN 978-981-10-7151-5 ISBN 978-981-10-7153-9 (eBook)
https://doi.org/10.1007/978-981-10-7153-9

Library of Congress Control Number: 2017957695

Printed on acid-free paper

This Springer imprint is published by Springer Nature
The registered company is Springer Nature Singapore Pte Ltd.
The registered company address is: 152 Beach Road, #21-01/04 Gateway East, Singapore 189721, Singapore

Contents

Prior-to-school	**Preschool**	**Early years of school**	**Early Childhood**	**Play**	**Everyday experiences and knowledge**	**Prior knowledge and mathematics**	**Funds of Knowledge**	**Home**
2, 10	3, 13	4, 6, 8, 10, 11, 14	2, 4, 5, 7, 12	2, 3, 4, 7, 9, 13	2, 3, 7, 8, 11, 13, 14	2, 4, 8, 10, 11, 12, 14	10, 13	7, 9, 10, 12, 13
Mathematical or statistical Reasoning	**Mathematical language**	**Mathematical concepts**	**Mathematical processes**	**Mathematical Structure**	**Environmental and task design**	**Curriculum**	**Pedagogy**	**Learning trajectories**
3, 5, 7, 11, 12, 11, 14	5, 6, 7, 8, 12, 13	2, 3, 4, 5, 6, 7, 14	4, 5, 6, 8, 9, 11, 14	2, 3, 4, 6, 11, 14	2, 3, 6, 7, 8, 11	2, 7, 10, 12, 13, 14	2, 3, 4, 7, 8, 13,	6, 7, 13
Number	**Counting**	**Geometry**	**Visualization**	**Gesture**	**Drawing**	**Objects and Materials**	**Children's Graphic Representations**	**Mathematizing**
3, 4, 6, 7, 8, 10, 14	6, 8, 14	5, 7	5, 7, 8	4, 5, 7, 8	4, 5, 11, 13	2, 3, 4, 5, 7, 8, 14	4, 5, 8, 11, 13, 14	4, 6, 8, 12

About the Editors

Dr. Virginia Kinnear is a Lecturer in early childhood education at Flinders University in South Australia. She worked as a solicitor for 10 years before coming to early childhood education through studying and teaching Montessori education (3–6 years) in the USA. She then qualified and taught in early childhood teaching qualification in Australia and has 18-years experience in early childhood and tertiary teaching. Virginia's Ph.D. focused on statistical reasoning in early childhood however her research and teaching interests are in all aspects of mathematics learning. She is particularly interested in young children's statistical learning and thinking, the development of critical and ethical thinking, and the intellectual attributes and processes that support mathematical engagement.

Dr. Mun Yee Lai is a Senior Lecturer in Mathematics at the Australian Catholic University. She has extensive experience and lectures in both primary and secondary mathematics education. She conducts research into pre-service teachers' mathematical knowledge for teaching, teaching for mathematising, using Variation Theory in Learning Study and Chinese primary children's misconception in decimal numbers. Recently, she has begun to explore the application of Planned Behaviour Theory for understanding school learners' mathematics learning intention.

Dr. Tracey Muir is a Senior Lecturer in Mathematics Education at the University of Tasmania. Her research interests include effective teaching for numeracy, problem solving in mathematics, the use of ICT in mathematics and parental involvement in mathematics education. She has published over 40 journal articles and conference papers and was a previous co-editor of Australian Primary Mathematics Classroom. She is a past Vice President (Development) of Mathematics Education Research Group of Australasia (MERGA). She has presented seminars and workshops at state, national and international conferences and is particularly passionate about working with teachers to enhance their classroom numeracy practices and engaging children in mathematics.

Chapter 1
Forging Connections in Early Mathematics: Perspectives and Provocations

Virginia Kinnear, Mun Yee Lai and Tracey Muir

Abstract This edited book brings together an international collection of work on a consistent and growing focus in mathematics education: the need to forge connections in early mathematics learning. Each chapter examines diverse ways that connections can be made, philosophically, theoretically, and pedagogically, illustrating different perspectives and providing provocations for researchers and educators. This chapter introduces themes of connection found in mathematics education and interdisciplinary literature and considers the purpose and value of connection in mathematics learning, specifically for children from birth to eight years. An overview of each of the book chapters and their connections mathematics is also provided.

Keywords Connections · Mathematical learning · Purposeful · Mathematics ideas Children's lives · Mathematics culture

1.1 Introduction

> Connection: A relationship in which a person or thing is linked or associated with something else; the action of linking one thing with another (Oxford Dictionary, 2017)

The significance of forging connections in young children's mathematical learning has gained increasing attention in the last few decades. Multidisciplinary research continues to expand our evidence for, and understanding of, young

V. Kinnear (✉)
Flinders University, Adelaide, Australia
e-mail: virginia.kinnear@flinders.edu.au

M. Y. Lai
Australian Catholic University, Sydney, Australia
e-mail: MunYee.Lai@acu.edu.au

T. Muir
University of Tasmania, Hobart, Australia
e-mail: tracey.muir@utas.edu.au

V. Kinnear et al. (eds.), *Forging Connections in Early Mathematics Teaching and Learning*, Early Mathematics Learning and Development,
https://doi.org/10.1007/978-981-10-7153-9_1

children's capacities to learn mathematics, and to widen our perspectives about providing children with rich, meaningful mathematics learning experiences. We also know that successful mathematics learning experiences significantly impact children's current and future education and life opportunities (Clements & Sarama, 2014; Moss, Bruce, & Bobis, 2016; Perry & Dockett, 2008). Research has provided researchers and educators with a growing awareness of children's aptitude for mathematical learning and provided grounds for arguing that we have strong ethical obligations to provide mathematically purposeful environments where young children can make sense of mathematics and develop their mathematical thinking.

What such environments should look like, what mathematics should be learned (and therefore taught), and how it should be taught, has stimulated both research and debate. The rapid development, changing capacities and the diversity of experiences for children aged birth to eight years of age present unique opportunities and challenges for teaching and learning. We continue to find connections between research and practice that map networks of knowledge about young children's mathematics learning. These in turn grow our understanding of the impact of learning connections made and missed both in the moment and in the future. The principal aim of this book has been to explore the common threads that are visible when researchers, educators, and children engage in forging connections in mathematical teaching and learning in seemingly disparate ways.

1.2 Connecting Young Children's Mathematical Learning

The concept of *connecting* is endemic in the burgeoning mathematics research of the last fifty years, indicative of the efforts being made to understand the complexities that characterize mathematics teaching and learning, and resulting in a research terrain that is 'extensive and diverse' (Kilpatrick, 2014, p. 271). 'Connections' have become synonymous with mathematical understanding, captured in Skemp's (1978) theory of relational learning, developed as relationships are forged across and between a myriad of networks. A sweep of theoretical models of mathematical learning repeatedly endorses a common theme: The promotion of mathematical understanding is achieved through children's ability to make cognitive connections to a network of previously connected experiences in multiple ways and their ability to apply interconnected knowledge to new mathematical situations (e.g., Clements & Sarama, 2009; Freudenthal, 1973; Haylock & Cockburn, 2008; Kaput, 1991; Lesh & Doerr, 2003; Lehrer & Schauble, 2000). A concept inherent in this approach is that mathematics itself provides structures that enable connections to be made.

The idea that the discipline of mathematics consists of repeated, interconnecting concepts is not new. Hardy (1940) stated that significant mathematical ideas are those that can be connected with a large complex of other mathematical ideas (p. 16). Steen (1990) defined mathematics as "about pattern and order of all sorts"

(p. 2). 'Deep ideas' found in mathematical structure, such as attributes, actions, abstractions and dispositions Steen argued, should be connected to children's educational mathematical experiences in ways that are appropriate to their own lives. The notion of a structural frame of big ideas' or 'powerful ideas' as central to mathematics and essential for mathematical sense-making is found in literature, policy, and curriculum (Clarke, Cheeseman, & Clarke, 2006; Early Years Learning Framework (DEEWR, 2009; National Council of Teachers of Mathematics (NCTM), 2000; Perry & Dockett, 2012). Although what defines these ideas is not definitive, they are seen as connectors 'between and within mathematical concepts and actions' (AAMT, 2009, p. 4). The current work of Mulligan and Mitchelmore (2016) continues to highlight that educational attention and connection to mathematical pattern and structure is pivotal to young children's mathematical learning.

As a result of this research, we have a better understanding of experiences that positively and negatively impact children's mathematical learning and therefore a broader understanding of the connections educators need to be mindful of and work to build. Coherence between an educator's network of mathematical connections and pedagogical actions he or she takes was found to support children to make their own connections to mathematical ideas and methods (Askew, Rhodes, Brown, William, & Johnson, 1997). We know the importance of the reciprocal relationship between connected mathematical knowledge in teaching and learning. This connecting knowledge has continued to grow, visible in the seminal work of Shulman (1986), Ball (1990), Ball, Thames, and Phelps (2008), and Rowland, Turner, Thwaites, and Huckstep (2009). Rowland et al.'s Knowledge Quartet pays particular attention to 'making connections,' emphasizing the logical and coherent nature of mathematics as a discipline, the need for pedagogy to connect to its concepts and procedures and for planned new learning to be synchronized and linked to children's former learning. For very young children, this entails educators working to connect their informal and intuitive knowledge to new mathematical experiences and new knowledge.

For children in prior-to-school settings, it can be particularly challenging to make connections to mathematics in environments where learning can be philosophically and theoretically misaligned with school-based settings. Conversely, for children in the early years of school, challenges lie in negotiating the theoretical and pedagogical approaches to teaching increasingly complex mathematics. Children from birth to 8 years experience multiple transitions as a result of the grouping children by age, creating complexities that are spotlighted in the critical work on children's transitioning to school. Characteristics of transition including opportunities, aspirations, expectations, and entitlements were identified by the Transition to School Position Statement (Educational Transitions and Change Research Group (ETC), 2011). As our colleagues in the first book in this series reminded us, mathematics education plays a major role in children's transition as they move into their first year of school. It is worth considering here too that these characteristics accompany times of change in young children's lives across. All children from birth to eight years face the challenge of their existing mathematical knowledge, ability, and potential being recognized in educational environments, and require engaging activities that advance their mathematical development with integrity and dignity. Forging connections to young children's cultural, social, emotional, and neurological dimensions of learning therefore take on particular importance.

1.3 The Need for Purposeful Connections

Under the UN Convention on the Rights of the Child (UNESCO, 1989), young children are deemed to have a right to education. The importance of children's development in the first few years of life is widely recognized and internationally promoted (United Nations Educational, Scientific and Cultural Organization (UNESCO), 2017). As a consequence, research evidence as to the value the early years hold for children's overall development and well-being and research into the role of families in children's early learning has also gained momentum (Phillipson, Gervasoni, & Sullivan, 2016). Further support is found in interdisciplinary research such as neuroscience that highlights the connections forged between children's sensory experiences and their physical and social experiences in particular. We are therefore aware of the critical function played by connections between environment and children's interaction with it, particularly in the development of neurological pathways (McCain, Mustard, & McCuaig, 2011). Psychology research further supports theories that young children's actions are focused on sourcing physical and social structure, that is, children are seeking connections in order to make sense and find meaning (Sommer, 2012). From this, we can infer that young children are purposeful in their search for meaning.

The purposeful role of children's actions in their sense-making underpins current theoretical explanations of mathematical learning, including constructivism and social constructivism. Principles found in Clements and Sarama's (2009, 2014) 'Learning Trajectory' work, for example, propose networks of mathematical learning with synergies and interplays between cognitive and environmental elements including sequenced instructional tasks, hierarchical progressions in mathematical developmental, and cultural connections. English (2004) emphasizes the need for children to be supported in establishing relational connections between various mathematical representations that are central to success in analogical and mathematical reasoning. Reiterating this in 2016, English (2016) refocuses on pedagogical connections needed in problem posing and solving practices. Such pedagogical practices can enable a balance to be found between cognitive demand and capacity, using approaches that are mindful of children's existing knowledge and context of understanding. In these ways, we can not only acknowledge young children's existing mathematical competence, but also find ways to 'bootstrap' and construct conceptual connections so that learning potential can be surfaced.

Common features of programs shown to be effective in supporting young children's access to mathematical ideas connect with children's current mathematical knowledge and interests, extend children's play into planned mathematical experiences, incorporate activities that are cognizant of children's development, and importantly, connect children with a competent adult (Moss et al., 2016). What is clear from the current mathematics education research is that young children's mathematical competency is consistently underestimated, with valuable learning opportunities unrealized as a result; none more so than for children aged birth to three years.

Recent reviews of mathematics education research highlight that the focus in early years mathematics literature is in preschool and the early years of school, capturing children aged from approximately four to eight years (e.g., Moss et al., 2016) and as a global community we have gained a great deal of knowledge in recent decades. Mathematics education research on the mathematical learning and teaching of very young children aged birth to three years is less prominent. There are significant numbers of studies on the mathematical learning of very young children; however, these are found predominantly in the disciplines of psychology and educational neuroscience. These studies provide important research evidence about the complexity and structures of the brain and their relationship to learning (e.g., van der van der Meulen, Krabbendam, & de Ruyter, 2015); however, the connections between this research and educational practice appears troubled from both sides. Fischer, Goswami, and Geake (2010) argue that neurological research should be both scientifically sound and connected to teaching and learning in educational settings to be of educational use in practice and policy. What can be added to this is that mathematics education research needs to be as serious in connecting to existing research in these fields. A great deal more rigorous debate is needed about the purpose of mathematical leaning from birth and how this can inform both research and practice. Empirical research is needed to develop a body of sound evidence about what can and should be learned, and what pedagogical practices are possible for relevant and meaningful mathematical learning to be inclusive of children aged birth to three years.

1.4 Disciplinary and Interdisciplinary Gaps and Disconnections

There is a schism between knowledge of the body and brain's role in children's processing and use of mathematics gained from research in neuroscience, psychology and related fields, and the paucity of pedagogical knowledge from mathematics education research as to how we can provide worthwhile opportunities for young children to learn mathematics. Education researchers are already moving, investigating transdisciplinary gaps in knowledge in specific areas of mathematics learning (see Bruce et al., 2016). The schism extends to the philosophical and arguably political divide that appears in response to seeing mathematics learning as a continuum that begins at birth. If mathematics learning begins at birth, what should mathematics teaching look like and who are children's teachers? The first book in this series investigated the transitioning children experience in their mathematics learning between prior-to-school and school learning (Perry, McDonald, & Gervasoni, 2015). It raised critical issues on the very real impact transitioning to school has on all participants in the process, which in turn raises questions about the mathematical learning children experience across the prior-to-school years, including very young children. Can all periods of a child's life in their early years be informed by a close examination of opportunities, entitlements, aspirations, and expectations? How do we approach a child's right to mathematics education with integrity and substance?

This book, the second book in the Springer Early Mathematics Learning and Development series, aims in part to challenge a reductive perspective on the value of the mathematical teaching and learning for children from birth to eight years, and specifically from birth to three years, a view that has historically pervaded mathematics education research. We urge mathematics education researchers to forge reciprocal interdisciplinary and cross-disciplinary research connections in order to find ways to connect to the mathematical learning needs of all children aged birth to eight years. This book connects to the first edited by Perry et al. (2015). Their focus on the theoretical and practical considerations implications of children mathematics learning in transitioning to school highlighted the importance of children's prior-to-school mathematics learning, particularly the role of home and community.

1.5 Using This Book: Forging Connections

1.5.1 Terminology: Naming the Early Years

Early childhood as defined by the Organization for Economic Co-operation and Development (OECD) (2001) and UNESCO (2017) is persons aged between birth and eight years. The early childhood sector is inevitably organized around children's ages, more commonly birth to age two (babies and toddlers), three to five years (before formal schooling commences), and five to eight years (the first three years of formal schooling). The literature is replete with multiple descriptions that reflect this, and so for language consistency, the book chapters and their keywords use the following terms to describe early childhood organizing structures:

- *young children*, *early years,* and *early childhood*—birth to eight years
- *prior-to-school*—prior to formal schooling, from birth to age five years
- *very young children*—birth to two years
- *preschool*—three years to school entry age, from three to five years

- *early years of school*—the first three years of formal schooling from five to eight years

1.5.2 Approaching the Book

This book aimed to weave threads of connections through disparate chapters, and the authors were asked to consider the ways their work, and so their chapters, could contribute to this. The two Contents pages provide two ways to approach reading the book. The first lists the chapters in alphabetical order by the last name of the first author and the set of keywords for that chapter. The chapter title and keywords provide insight into the identified connections *within* the chapter. The second provides a table that connects the chapters together using the keywords. This enables the reader to look at a keyword and see which chapters have connections to that idea, highlighting the connections *between* chapters. We pick up these chapter connections and expand on these in the next sections of this chapter.

We hope that by highlighting different perspectives on unifying themes that the central tenet of the book will be visible: that there is important work to be done in forging connections in young children's mathematical learning. In calling for chapters, we were struck by how little research is currently undertaken into mathematics learning for children aged birth to three years. We were struck also by authors' reactions when approached to contribute to the book as many were surprised, and respectfully but firmly of the view, that a contribution was not possible as 'mathematics isn't my area.' These responses belie the very real and important opportunities for interdisciplinary collaboration and for attention to be focused on the mathematics learning experiences of all young children from multiple perspectives. We need to be thinking about the mathematical experiences young children can access in the moment and also how we can plan to extend their mathematical horizons (Ball et al., 2008).

The choices educators make about the mathematical content, processes and dispositions they choose to foster in early learning environments are important, and have very real impact on children's immediate and long-term future. With these aspects in mind, we highlight some of the significant connections in young children's mathematical teaching and learning that are foregrounded in the book. Chapter connections that relate to the ideas are in brackets, but we ask that you consider the connections as common threads that should be visible in both research and teaching.

1.6 Forging Connections with Children's Lives and Worlds

Recent research in early childhood mathematics education in Australasia reflects the movement toward an examination of the environmental, relational, and sociocultural practices around mathematics (MacDonald, Goff, Dockett, & Perry, 2016). It has been well documented that the environmental context of early childhood mathematics education extends beyond the physical classroom or child center and into the sites where children live and learn (Perry, Gervasoni, & Dockett, 2012), factors that can be predictive of children's mathematical achievement (Chap. 12). It is therefore important to consider the mathematical capital that young children possess before beginning formal education, including acknowledging the 'broad range of formal mathematics knowledge that many children construct' (Perry, 2013, p. 344) in home and preschool contexts, and children's interests in and capacity to engage with mathematical ideas (Chap. 2). Connecting children's early and varied mathematical representations to mathematical concepts (Chaps. 4 and 8) and recognizing the mathematical capacity children bring to mathematical judgments (Chaps. 11 and 14) can enable educators to connect to children's existing knowledge and extend their mathematical experiences in prior-to-school, preschool, and the early years of school settings.

In addition to recognizing the social and cultural capital that young children bring with them into more formal educational contexts, parents and families should also be viewed as a source of capital and acknowledged for the contributions they can and do make to their child's education (Chap. 10) and form part of the continuum of children's mathematical learning (Chaps. 7 and 14). Research has shown that there are a number of positive consequences associated with family involvement in a child's education, including improving levels of student health, increasing academic achievement, fostering positive attitudes toward learning and school, and increasing parent–child interaction (Peressini, 2000). Despite the benefits, there is widespread evidence to show that parents often feel uninformed about contemporary mathematical educational practices and limited in their ability to notice and support mathematical learning at home (e.g., Muir, 2012; Pritchard, 2004). Other research, however, shows that parents are keen to encourage and support their children in their mathematics education, including those from low SES and culturally different backgrounds (e.g., Anthony & Walshaw, 2007; Muir, 2009).

Mathematical practices and children's engagement flow from the decisions that are made by adults and educators about the purpose, content, and processes of the experiences they offer. We ask then, what concepts and processes should be foregrounded for successful learning? How are we to connect to what children already have by way of mathematical interests and competencies? (Chaps. 3, 7, 9, 10, 13, and 14). What do we notice about these and the cultural knowledge that underpins their mathematical play and engagement with mathematical concepts? 'Funds of knowledge,' (attributed to Moll, Amanti, Neff, & Gonzalez, 1992) is defined as 'the historically accumulated and culturally developed bodies of knowledge and skills essential for household or individual functioning and well-being' (p. 133). When educators access this knowledge (Chap. 9), they are better equipped to provide culturally responsive and meaningful experiences for children that tap into their prior knowledge (Lopez, 2006). Increasing involvement of parents in partnerships with educators can connect children's mathematical experiences at school to home. Successful initiatives to forge these connections can increase parental understanding of contemporary mathematics education practices and strengthen the cultural relationships between home and educational environment (Chap. 9).

Play can provide powerful insights into children's mathematical knowledge, and while informal mathematics may differ considerably from standard 'school' mathematics, though children's cultural knowledge and mathematical texts can reveal considerable mathematical understandings (Chap. 13). Educators play an important role in building connections between young children's representations and their conceptual development in mathematics through play-based activities. The structure and design of mathematical activities, and how they direct children's attention to mathematical content can capitalize on children's learning from everyday experiences (Chaps. 3, 8, and 14). Mathematical connections can be found in young children's experiences that are accompanied by mathematically generated representations, including language, gesture, and the context of play, and these can yield helpful insights into their mathematical thinking and reasoning and play an important role in children's mathematical learning (Chaps. 4, 7, and 10). When considering children's play-based experiences and everyday experiences, attention should be given to the mathematical authenticity of children's mathematical activity that such experiences provide (Chap. 2).

1.7 Forging Connections to Mathematics History and Culture

Realistic Mathematics Education [RME] and the work of Freudenthal (1905–1990) and colleagues (see for example Van den Heuvel-Panhuizen, 2001) have fueled and influenced mathematical debate, research, and practice for decades. As a reform approach to mathematics education, RME emphasizes conceptual understanding and concept development in mathematical learning. The core of the approach is *mathematizing* that focuses attention on how realistic situations that can be used to generate mathematical learning and develop a series of connections between concrete and abstract conceptual mathematical models. RME was generated by rebellion against mathematics being taught as a disconnected, systematic 'product' discipline that children 'received.' Instead, mathematics was taken as first and foremost a human activity, meaning that mathematical learning should be an active experience where real-life problems require mathematical concepts and strategies to be organized and mathematized in order for the problem to be solved (Van den

Heuvel-Panhuizen, 2001). The mathematization process therefore aims to provide a 'guided reinvention' of mathematical development that children can access in order to make meaning of experiences mathematically (Chaps. 4, 6, 8, and 12). This in turn enables the integrated nature of mathematical relationships to be revealed and refined as children develop conceptual models that bridge informal and formal mathematics and share strategies and solutions (Van den Heuvel-Panhuizen, 2001).

Mathematizing reflects a move to a more holistic approach to teaching and learning that is relevant for early childhood. The process of mathematizing is a direct response to experiences that disconnect mathematical learning from mathematical processes, distorting the value and purpose of mathematics. How can we engage children in the exploration and experience of the evolution of mathematics concepts? How can experiences be designed that can provide conceptual bridges between concrete mathematical experience and more formal generalized mathematical concepts, symbols, and signs? (Chaps. 3, 5, 6, 7, and 8). At the heart of mathematizing is mathematics that emerges from children's interaction with their environment, where children are encouraged to develop their contextual tools for solving real-life mathematical problems (Chaps. 4, 5, 6, and 8). Experiences should enable children to have sustained engagement with mathematical concepts and build mathematical relationships over time. In highlighting an approach to how mathematics has come to be, mathematizing raises deeper questions about how children gain mathematical knowledge. We must then make pedagogical choices; what mathematics, taught when, and taught how (Chaps. 2, 3, 4, 7, 8, and 13).

1.8 Forging Connections to Mathematics Ideas

Early years mathematics research has historically been polarizing. Young children are seen either as limited mathematically (a focus on what they are unable to do) or as mathematically capable and therefore under challenged. These perspectives create both tension and opportunities: What are appropriate pedagogical practices that can recognize children's capacity, and enrich and extend children's existing mathematical knowledge without compromising their entitlement to playful education that foster well-being? Mathematical concept development can be supported through educators' choices of objects (Chaps. 2, 3, 4, 5, 7, 8, and 14), in the structure of teaching activities (Chaps. 2, 3, 6, 7, 8, and 11), and the children's attention and intention to learning (Chap. 2). The thoughtful and careful choices of narratives and materials educators' make are essential tools for teaching mathematics in the early years. They are tools for educators to connect children's daily experiences in informal mathematical knowledge to the construction of new mathematical concepts (Chaps. 6, 7, 13, and 14).

How we view mathematics has much to do with how we view ourselves as mathematics learners, educators, and researchers (Chap. 9). This in turn influences theoretical, content and pedagogical choices and decisions in research and educational settings (Chap. 2). Thinking about mathematics as an abstract structured discipline demands that we define what is at the core of the structure, particularly for young children who are building from the ground up. If, as many chapter authors have argued, mathematics is a human activity, socially developed and able to be applied in real and abstract ways, then we have a basis for determining how the core concepts, processes, and relationships can be identified, accessed, and learned. Children's activity with mathematics, that is, active experiences with core mathematical concepts shift attention from passive to active learning, one where knowledge is 'the productive achievement of the student' in interaction with others (Wittmann, 2005, p. 94). Learning in this frame requires activities where focused but flexible interaction with mathematical processes, procedures, and content is required through investigation of real or mathematical situations (Freudenthal, 1973). This challenges us to think about what mathematics is, and what mathematics *education* in early childhood is. We ask: What criteria do we use to determine the mathematical content of young children's mathematical experiences, what are 'big ideas' that influence this? (Chap. 7). Have we created an artificial divide between pure and applied mathematics in young children's mathematical learning? (Chap. 2). In what ways do our decisions on the design of children's mathematical experiences connect or disconnect children from access to the abstract structure and pattern of mathematics as a discipline? (Chaps. 2, 3, 5, 6, 11, and 14).

Abstract mathematical concepts or properties can be developed using mental and physical processes. Young children can access a range of ways to connect their learning to the abstract forms in measurement, geometry (Chaps. 5 and 7), number (Chaps. 3, 4, 6, 7, 8, 10, and 14), and data and statistics (Chap. 11). Powerful connectors are found in visualization (Chaps. 5, 7, and 8) and accompanying gesture (Chaps. 4, 5, 7, and 8), representations (Chaps. 4, 5, 8, 11, 13, and 14), language (Chaps. 5, 6, 7, 8, 12, and 13), and reasoning (Chaps. 3, 5, 7, 11, 12, 11, and 14). Tasks that stimulate deep spatial insights in children's early geometry learning and young children's ability to use visualization can connect across geometric concepts (Chap. 4). Spatial awareness and shape can connect and traverse the bridge between informal learning and teaching for geometric understanding (Chap. 6). Language, gesture, and visualization can act as mediators between concrete experiences and abstract thinking, and informal learning opportunities can be extended into challenging and relevant learning in home and educational environments (Chap. 7).

Realizing the mathematical potential in informal learning opportunities and planning for relevant mathematical tasks requires an alignment between the learning environment that is created and an educator's mathematical knowledge and understanding. Educators' mathematical identities and their perceived roles as mathematics educators impact developing relevant mathematical content and meaningful experiences for children (Chap. 8). Transformative change is possible when professional development challenges conceptions about mathematics and mathematical identity, which can in turn connect or disconnect educators to meaningful mathematical practices with children. Educators' theoretical knowledge can therefore be predictors for successful mathematical learning with young children (Chap. 11). To consider this is a salient reminder that young children's opportunities are wide-ranging in their type and impact and are predictive of children's ability to realize their mathematical learning potential. Mathematical learning and the role of interdisciplinary knowledge in teaching and learning are intimately connected. Mathematical knowledge and collaboration should guide early years' mathematics policy, practice, and professional development, and we argue, should underpin research in mathematics education in early childhood, particularly for children aged birth to three years.

1.9 A Final Note on Reading This Book

This is a book for reading in multiple ways, and to read with a view to making connections between your own knowledge and understanding and that which is offered in each chapter. To support your own connection making and to highlight the many ways that seemingly dissimilar chapters can connect, there are two contents pages. The first provides a list of the chapters, which are in alphabetical order by the last name of the first author. Next to each chapter listing is a list of keywords identified by the authors. The second contents page provides a grid that groups chapters under common theme keywords.

The concluding chapter in this book asks us to consider 'connection metaphors' in mathematics, along with the time it takes to develop strong connections and the mathematical purpose of the connections we seek. We hope you will be inspired to reflect on the connections you forge, whether in research or practice, and to consider new connecting possibilities.

References

Australian Association of Mathematics Teachers. (AAMT). (2009). *School mathematics for the 21st century: What should school mathematics of the 21st century be like?* Discussion Paper. Available: http://www.aamt.edu.au/Library/Other/Maths-for-the-21st-century.

Anthony, G., & Walshaw, M. (2007). *Effective pedagogy in mathematics/Pangarau*. Wellington, NZ: Ministry of Education.

Askew, M., Rhodes, V., Brown, M., William, D., & Johnson, D. (1997). *Effective teachers of numeracy: Report of a study carried out for the Teacher Training Agency*. London, Kings College: School of Education.

Ball, D. (1990). The mathematical understanding that prospective teachers bring to teacher education. *Elementary School Journal, 90*, 449–466.

Ball, D., Thames, M. H., & Phelps, G. (2008). Content knowledge for teaching: What makes it special? *Journal of Teacher Education, 59*(5), 389–407.

Bruce, C., Davis, B., Sinclair, N., McGarvey, L., Hallowell, D., Drefs, M., … Woolcott., G. (2016). Understanding gaps in research networks: Using "spatial reasoning" as a window into the importance of networked educational research. *Educational Studies in Mathematics*, *95*(2), 143–161. DOI 10.1007/s10649-016-9743-2.

Clarke, D. M., Cheeseman, J., & Clarke, B. (2006). The mathematical knowledge and understanding young children bring to school. *Mathematics Education Research Journal, 18* (1), 78–102.

Clements, D., & Sarama, J. (2009). *Learning and teaching early math: The learning trajectories approach*. New York, NY: Routledge.

Clements, D., & Sarama, J. (2014). *Learning and teaching early math: The learning trajectories approach*. New York, NY: Routledge.

Department of Education, Employment and Workplace Relations (DEEWR). (2009). *Belonging, being and becoming: The early years learning for Australia*. Canberra: Commonwealth of Australia.

Educational Transitions and Change (ETC) Research Group. (2011). *Transition to School: Position Statement*. Albury-Wodonga: Research Institute for Professional Practice, Learning and Education, Charles Sturt University. Available on-line: http://www.csu.edu.au/faculty/educat/edu/transitions/publications/Position-Statement.pdf.

English, L. (2004). Promoting the development of young chidlren's mathematical and analogical reasonng. In L. D. English (Ed.), *Mathematical and analogical reasoning of young learners*. Mahwah, NJ: Lawrence Erlbaum Associates.

English, L. (2016). Revealing and capitalising on young children's potential. *ZDM Mathematics Education, 48*(7), 1079–1087.

Fischer, K., Goswami, G., & Geake, J. (2010). The future of educational neuroscience. *Mind, Brain and Education, 4*(2), 68–80.

Freudenthal, H. (1973). *Mathematics as an educational task*. Dordrecht: Reidel.

Gifford, S. (2004). A new mathematics pedagogy for the early years: In search of principles for practice. *International Journal of Early years Education, 12*(2), 99–115.

Hardy, G. H. (1940). *A mathematician's apology*. Cambridge: Cambridge University Press.

Haylock, D., & Cockburn, A. (2008). *Understanding mathematics for young children: A guide for Foundation Stage and lower primary teachers*. London: Sage.

Kaput, J. J. (1991). Notations and representations as mediators of constructive processes. In E. von Glaserfeld (Ed.), *Constructivism in mathematics education* (pp. 53–73). Dordecht, Netherlands: Kluher.

Kilpatrick, J. (2014). History of research in mathematics education. *Encyclopedia of mathematics education* (pp. 267–272). Netherlands: Springer.

Lehrer, R., & Schauble, L. (2000). Modeling in mathematics and science. In R. Glaser (Ed.), *Advances in instructional psychology* (Vol. 5, pp. 101–159)., Educational design and cognitive science Mahwah, NJ: Lawrence Erlbaum.

Lesh, R. A., & Doerr, H. (Eds.). (2003). *Beyond constructivism: A models and modelling perspective on mathematics teaching, learning and problem solving*. Mahwah, NJ: Lawrence Erlbaum.

Lopez, J. K. (2006). Funds of knowledge. Available: http://www.learnnc.org/lp/editions/brdglangbarriers/939.

McCain, M. N., Mustard, J. F., & McCuaig, K. (2011). *Early Years Study 3: Making decisions, taking action*. Toronto: Margaret and Wallace McCain Family Foundation.

MacDonald, A., Goff, W., Dockett, W., & Perry, B. (2016). Mathematics education in the early years. In K. Makar, S. Dole, J. Visnovska, M. Goos, A. Bennison, & K. Fry (Eds.), *Research in mathematics education in Australasia 2012–2015* (pp. 165–186). Singapore: Springer.

Moss, J., Brice, C. D., & Bobis, J. (2016). Young children's access to powerful mathematics ideas. In L. English & D. Kirshner (Eds.), *Handbook of international research in mathematics education* (3rd ed., pp. 153–190). New York: Taylor & Francis.

Moll, C., Amanti, C., Neff, D., & Gonzalez, N. (1992). Funds of knowledge for teaching: Using a qualitative approach to connect homes and classrooms. *Theory Into Practice, 31*(2), 132–141.

Mulligan, M., & Mitchelmore, M. (2016). *PASMAP: pattern and structure mathematics awareness program*. Camberwell, Vic: ACER.

Muir, T. (2009). At home with numeracy: Empowering parents to be active participants in their child's numeracy development. In R. Hunter, B. Bicknell & T. Burgess (Eds.), *Crossing divides* (*Proceedings of the 32nd annual conference of the Mathematics Education Research Group of Australasia,* pp. 395–402). Wellington, NZ: MERGA.

Muir, T. (2012). It's in the bag: Parental involvement in a numeracy at-home program. *Australasian Journal of Early Childhood, 37*(2), 106–112.

National Council of Teachers of Mathematics (NCTM). (2000). *Principles and standards for school mathematics*. Reston, VA: Author.

Organisation for Economic Co-operation and Development (OECD). (2001). *Starting strong: Early childhood education and care*. Paris: Author.

Oxford Dictionary (2017). Retrieved from https://en.oxforddictionaries.com/definition/connection.

Peressini, D. D. (2000). What's all the fuss about involving parents in mathematics education? In D. Edge (Ed.), *Involving families in school mathematics* (pp. 5–10). Reston, VA: NCTM.

Perry, B. (2013). Preschool and school educators noticing young children's mathematics. In V. Steinle, L. Ball, & C. Bardini (Eds.), *Mathematics education: Yesterday, today and tomorrow* (Proceedings of the 36th annual conference of the Mathematics Education Research Group of Australasia, Melbourne, pp. 783–786). Melbourne: MERGA.

Perry, B., & Dockett, S. (2008). Young children's access to powerful mathematical ideas. In L. D. English (Ed.), *Handbook of international research in mathematics education* (2nd ed., pp. 75–108). New York: Routledge.

Perry, B., Gervasoni, A., & Dockett, S. (2012). Let's count: Evaluation of a pilot early mathematics program in low socioeconomic locations in Australia. In J. Dindyal, L. P. Cheng, & S. F. Ng (Eds.), *Mathematics education: Expanding horizons* (Proceedings of the 35th annual conference of the Mathematics Education Research Group of Australasia, Singapore, pp. 594–601). Singapore: MERGA.

Perry, B., MacDonald, A., & Gervasoni, A. (Eds.). (2015). *Mathematics and transition to school: International perspectives*. Singapore: Springer.

Phillipson, S., Gervasoni, A., & Sullivan, P. (Eds.). (2016). *Engaging families as children's first educators*. Singapore: Springer.

Pritchard, R. (2004). Investigating parental attitudes and beliefs in mathematics education. In I. Putt, R. Faragher & M. McLean (Eds.), *Mathematics education for the third millenium: Towards 2010* (Proceedings of the 27th Annual Conference of the Mathematics Education Research Group of Australasia, Townsville, pp. 478–485). Sydney: MERGA.

Rowland, T., Turner, F., Thwaites, A., & Huckstep, P. (2009). *Developing primary mathematics teaching*. London: Sage Publications.

Sarama, J., & Clements, D. (2009). *Early childhood mathematics education research: Learning trajectories for young children*. New York, NY: Routledge.

Shulman, L. (1986). Those who understand: Knowledge growth in teaching. *Educational Researcher, 15*(2), 4–14.

Skemp, R. (1978). Relational understanding and instrumental understanding. *Arithmetic Teacher, 26*(3), 9–15.

Sommer, D. (2012). *A childhood psychology: Young children in changing times*. Palgrave: Macmillan.

Steen, L. A. (1990). Pattern. In L. A. Steen (Ed.), *On the shoulders of giants: New approaches to numeracy* (pp. 1–12). Washington DC: Mathematical Sciences Board.

United Nations Educational, Scientific and Cultural Organization (UNESCO). (1989). *Conventions on the rights of the child*. New York: Author.

United Nations Educational, Scientific and Cultural Organization [UNESCO]. (2017). *Early childhood care and education.* Retrieved from: http://en.unesco.org/themes/early-childhood-care-and-education.

Van den Heuvel-Panhuizen, M. (2001). Realistic mathematics education in the Netherlands. In J. Anghileri (Ed.), *Principles and practices in arithmetic teaching: Innovative approaches for the primary classroom* (pp. 49–63). Buckingham: Open University Press.

van der Meulen, A., Krabbendam, L., & de Ruyter, D. (2015). Educational neuroscience: Its position, aims and expectations. *British Journal of Educational Studies, 63*(2), 229–243.

Wittmann, E. C. (2005). Mathematics as the science of learning from patterns: A guideline for developing mathematics education from early childhood to adulthood. *Pleany lecture at the International Colloquium 'Mathematical Learning from Early Childhood to Adulthood'*. Belgium, Mons.

Authors Biography

Dr. Virginia Kinnear is Lecturer in early childhood education at Flinders University in South Australia. She worked as a solicitor for 10 years before coming to early childhood education through studying and teaching Montessori education (3–6 years) in the USA. She then qualified and taught in early childhood teaching qualification in Australia and has 18-years experience in early childhood and tertiary teaching. Virginia's Ph.D. focused on statistical reasoning in early childhood and her research and teaching interests are in all aspects of mathematics learning. She is particularly interested in young children's statistical learning and thinking, the development of critical and ethical thinking, and the intellectual attributes and processes that support mathematical engagement.

Dr. Mun Yee Lai is a Senior Lecturer in Mathematics at the Australian Catholic University. She has extensive experience and lectures in both primary and secondary mathematics education. She conducts research into pre-service teachers' mathematical knowledge for teaching, teaching for mathematizing, using Variation Theory in Learning Study and Chinese primary children's misconception in decimal numbers. Recently, she has begun to explore the application of Planned Behavior Theory for understanding school learners' mathematics learning intention.

Dr. Tracey Muir is a Senior Lecturer in Mathematics Education at the University of Tasmania. Her research interests include effective teaching for numeracy, problem solving in mathematics, the use of ICT in mathematics, and parental involvement in mathematics education. She has published over 40 journal articles and conference papers and was a previous co-editor of Australian Primary Mathematics Classroom. She is a past Vice President (Development) of Mathematics Education Research Group of Australasia (MERGA). She has presented seminars and workshops at state, national, and international conferences and is particularly passionate about working with teachers to enhance their classroom numeracy practices and engaging children in mathematics.

Chapter 2
Early Mathematics Education: A Plea for Mathematically Founded Conceptions

Virginia Kinnear and Erich Ch. Wittmann

Abstract What is mathematics and what is mathematics education in early childhood? This theoretical chapter draws attention to distinctions and relationships between applied and pure mathematics and provides a provocation for considering what each offers when evaluating the authenticity of mathematical activity. The implications of these differences in how mathematics education are positioned and framed in early childhood are discussed using a number of examples to illustrate this. The chapter challenges perceptions by educators that mathematical experiences for prior-to-school learning must be contextualized in 'everyday' experiences and offers other ways of connecting young children's mathematical education to mathematics as a system with structure and abstract forms and patterns. The chapter draws attention to and pleads for mathematically founded conceptions to underpin the context and content of early childhood mathematics education.

Keywords Mathematical structure · Conceptual understanding
Design · Mathematics activities · Play · Pure and applied mathematics
Early childhood

2.1 Introduction—Corruptio Optimi Pessima (The Corruption of the Best Is the Worst)

When Friedrich Froebel founded kindergartens 170 years ago, he positioned mathematics as integral to all learning, as it 'is the expression of life as such: therefore its nature may be studies in life, and life may be studied with its help' (2005, p. 206). His deliberate integration of elements of mathematics into

V. Kinnear (✉)
Flinders University, Adelaide, Australia
e-mail: virginia.kinnear@flinders.edu.au

E. Ch.Wittmann
TU Dortmund University, Dortmund, Germany
e-mail: wittmann@math.tu-dortmund.de

V. Kinnear et al. (eds.), *Forging Connections in Early Mathematics Teaching and Learning*, Early Mathematics Learning and Development,
https://doi.org/10.1007/978-981-10-7153-9_2

educators' work with children is visibly demonstrated by his geometric 'gifts' and 'occupations,' which embody his philosophy that mathematics learning should be physical and dynamic. Froebel's gifts are a curriculum of exercises using specifically designed materials for children to work with. The gifts were structured to systematically develop, and logically connect knowledge of fundamental mathematical forms by drawing attention to the attributes and qualities of the objects children worked with. Beliefs that young children are both interested in mathematics and have capacity for mathematics learning are experiencing a resurgence in international policy, curriculum and practice, with increasing recognition of, and attention to the importance of early mathematical education (see McDonald, Goff, Dockett, & Perry, 2016, for a recent review of Australasian early childhood mathematics research). Such attention is found even in Germany where the ruling philosophy in early education over the past forty years has explicitly excluded any connections with school subjects. In Australia, the prior-to-school curriculum, the Early Years Learning Framework, captures mathematics education under a numeracy umbrella and subsumes it into an effective communication learning outcome along with literacy (DEEWR, 2009). The increased interest in young children's mathematical learning, and hence teaching, has prompted the authors to consider the purpose and substance of young children's mathematics education and the consequences that must follow an educator's decisions about these, that is, the mathematical learning context and content that young children are offered. In doing so, the authors attempt to delineate a position which enables mathematics educators to evaluate the mathematical substance of materials that are offered to children and the implications of these evaluations for early childhood mathematics education.

The renewed interest in mathematical education for young children is a positive sign, however, it does not by itself guarantee that the development of theoretical and pedagogical knowledge and understanding will move in a positive direction; one that will recognize children's existing potential and foster their early mathematical learning through mathematically informed and structured experiences. On the contrary, disciplines that have the biggest impact on our understanding of early mathematical learning, such as psychology, neuroscience, health and sociology, have provided evidence and fueled varied views about the why, what, and how of mathematical learning in both public policy and public perception. Without a critical perspective across the current range of available knowledge about mathematics teaching and learning there is the very real potential for misrepresentation or for misleading perspectives to gain favor in education. Conceptions of early mathematics and mathematical learning, such as 'developmentally appropriate, 'child-centered,' 'playful' and 'active' are widely found in current mathematics education theory and have value theoretically and pedagogically for children's learning. What we argue is that on their own, these conceptions do not have an inherent perception of mathematics as a discipline. As a result, their integration into mathematical learning and teaching are not inherently cognizant of how the discipline of mathematics is represented in pedagogical practice, and so distort its educative purpose and possibilities. For these reasons, we argue, therefore, any educational decision that does not critically evaluate the *mathematical* purpose and *mathematical concepts* at the foundation of early childhood mathematics educational experiences are not in children's best long-term interests. We are troubling the very purpose of early mathematics education.

2.2 Placing Mathematics Knowledge at the Core of Early Mathematics Education

Increased interest in young children's mathematics learning has coincided with a renewed interest in what constitutes knowledge and about the purpose of education. The work of Young (2010a, 2010b) and Young and Muller (2010, 2013, 2016), for example, considers the current tensions between knowledge, theory, policy, and education. They draw attention to disciplinary differences in internal coherence, concepts, content, and skills, and the ways different disciplines reach agreement about how reliable, stable objective knowledge (truth) is achieved and shared. From this perspective, a discipline's structure reveals 'hierarchies of abstraction' formed from conceptual development and maturation. Mathematics exemplifies a discipline with 'long sequences of hierarchically-related concepts' (Young & Muller, 2010, p. 21). There are reasons therefore to ask that we reconsider the role of specialist disciplinary knowledge in education. Such knowledge Young and Muller argue is not accessible or developed through informal experiences alone. Knowledge instead emerges from but is 'not reducible to the context in which it is produced and acquired' (2016, p. 67). This is a leading point as it highlights the role of the learning experience in the way knowledge is acquired, but also raises questions about the objectivity of the disciplinary knowledge that is engaged in the learning experience. How then should we evaluate the disciplinary knowledge that should be embedded and emerge from a mathematical learning experience?

If we consider mathematics education to be a design science (Wittmann, 1998, 2005), there are synergies with the perspectives provided by Young and Muller. Both perspectives position mathematics as a discipline with unique qualities, processes, and aims, which if placed at the core of mathematics education generate consequences for how we see mathematics itself, and therefore how we connect mathematics with mathematics education. Wittmann has long argued that it is only when mathematics education researchers' work with the disciplinary specific structure, content and purposes of mathematics, that what is needed to design learning environments that will engage children in mathematics learning can be uncovered and developed. This approach positions the dynamic but cohesive structure of mathematics as the driver of mathematical processes and procedures that can frame the context in which they are developed. It is the structure and the accompanying attributes that are connected to mathematics we argue, that should form the substance of a mathematical learning environment for young children. In a situation where mathematics is placed at the core of the learning experience, it is crucial to adopt a critical position using mathematical criteria. With mathematics at the core of analysis, it is possible to 'sort the wheat from the chaff' when considering the materials and types of experiences offered to children in early mathematics education environments, and to organize both pre-service and in-service education of early childhood educators in a way that takes such perspectives into account.

Accordingly, the following sections consist of two parts: First, some typical examples from various early mathematics programs will be critically analyzed, using mathematical criteria. The second part is a plea for a theoretical approach to early mathematical education, a view that challenges those that postulate that early mathematics learning is marked only by 'learning with all senses' and engaging 'everyday mathematics.'

2.3 Mathematical Criteria for Evaluating Early Mathematics Programs

Jean Piaget's research into the development of the concepts of number and space has stimulated a great variety of empirical studies. Mathematics researchers and educators have extended, refined, and modified Piaget's findings and elaborated on broader theoretical underpinnings. Particular emphasis in these and other studies concerned with children's prior-to-school experiences was placed on determining the prerequisite knowledge children bring to school (e.g., the review in Moss, Bruce, & Bobis, 2016; and also Clarke, Clarke, Grüßing, & Peter-Koop, 2008; Hengartner & Röthlisberger, 1995; Krajewski, 2008; Mulligan, Mitchelmore, English, & Crevenston, 2013; Perry & Dockett, 2008; Sarama & Clements, 2009; Schmidt & Weiser, 1982; Van den Heuvel-Panhuizen, 1995; Wittmann, 2004). The conclusions from these studies are unambiguous:

1. Prior-to-entering school, children are mathematically active and in general bring a great deal more mathematical knowledge to school than teachers are aware of which has developed from their everyday experiences.
2. The extent to which children are capable of unfolding and developing their mathematical potential depends on their social environment. In unfavorable conditions, there are dual impacting factors as children may have less prerequisite knowledge available and less opportunity to develop existing knowledge. Therefore, early mathematics education is crucial for children who do not receive sufficient or relevant support in their social environment, in particular children with disadvantaged backgrounds. This position is supported by research that considers the impact of program quality on children's cognitive development (Krieg, Curtis, Hall, & Westenburg, 2015; Sylva, Melhuish, Sammons, Siraj-Blatchford, & Taggart, 2004).
3. Properly conducted early mathematics education does have significant, very positive effects on the long-term development of mathematical thinking and later academic success, and is therefore also an equity issue.

Although these empirical studies provide us with good reasons for establishing mathematics education in early childhood, they do not provide us with criteria for evaluating the mathematical quality of early childhood mathematics programs. Research has raised more broadly the profile of early childhood mathematics planning and programming and begun to identify attributes that contribute to a 'quality' program. Such attributes include determining learning aims and pedagogical choices educators' make, choices that are critically important to children's short- and-long term learning outcomes (Sylva et al., 2004). Sylva et al.'s study, for example, draws attention to educator qualifications and experience in disciplinary learning areas such as mathematics. This knowledge is instrumental in the inclusion of teaching experiences that intentionally use cognitively challenging tasks to encourage children to question and share their mathematical thinking over time. The mathematical aspects of these experiences, however, were not analyzed. It is not by accident that studies that are conducted, including those by psychologists, do not often pay enough or appropriate attention to the mathematical aspects of mathematics teaching and learning. A typical example is Pauen (2009) in which the tests used did not reveal differences between two programs that differed greatly in the mathematical perspectives they held and used. With this in mind, we now consider ten examples of mathematics experiences.

2.4 Ten Examples

This section presents ten examples of mathematics materials and evaluates their mathematical value. This analysis will be a precursor to considering criteria for evaluating early childhood mathematics programs that maintain the integrity of mathematics as a discipline.

(1) *Wickie and the Strong Men* (German translation: 'Wickie und die starken Männer' (Anonymous authors, 2006)

Vicke the Viking ('Wickie' in German), a child character who relies on his brain to solve problems, is from a children's book series by Swedish author Jonsson published between 1963 and 1993 and popularized through cartoon and film adaptations. Similar to many publications that use popular children's characters, Wickie is central to a book of mathematics activities based on the various media versions of the stories that asks children to 'search, find, connect, count, and assign.' The figure 'Wickie,' well-known to many European children from a television series, acts as a guide through the activities. A typical example is devoted to the concepts of 'left' and 'right.' On the page the outline shapes of elks are visible. The text, using the character Halvar, Wickie's father (who usually uses muscles rather than thinking to solve problems) reads: 'How annoying! Halvar always forgets where the left and the right side is. This has to be changed. Please color red all elks moving to the left side. Color green all elks moving to the right side. Then Halvar should have less problems in distinguishing left and right.'

(2) *Make a bigger puddle, make a smaller worm* (Walter, 1972)

This book, also published as ‘The Magic Mirror Book’ (1984), was devised and illustrated by Marion Walter, an American mathematics educator. Each page shows an animal or an object that can be manipulated using a mirror, so that both the picture in the book together with its mirror image forms a new picture. For example, a picture of a worm in the book can be made shorter or longer depending on where the mirror is placed on the image. Similarly, a puddle can be bigger and smaller or a laundry line shorter and longer. The idea is continued with more complexity in Walter’s ‘Mirror Puzzle Book’ (1985), which requires children to work out where to place a mirror to match images in the book, and in Hartmut Spiegel’s (1997) ‘mirror cards’ where it is used to investigate symmetry and spatial relationships.

(3) *Zahlenland [Land of numbers]* (Preiß, 2004)

As the title suggests, this program is mainly devoted to the development of number concepts. Numerical figures are represented as characters that act and speak like human beings: ‘Good morning, dear numbers!’ is the title catchphrase for the program. To work with this program, an empty room is used to build a village with houses in which the numbers live, and a street which connects the houses. There are also ‘mistake devils’ and a ‘mistake forest.’ Children perform activities that are stimulated by stories. These stories consciously rely on children’s perceived preference for fantasy worlds and play that is connected to those worlds. For example, in one story the number 4 is ill and lying in bed. Numbers 1, 2, and 3 visit 4 and heal it in the following way: number 1 brings one heart as a present, number 2 rings two bells, and number 3 administers three drops of a blue medicine.

(4) *Mathe Kings [Math Kings]* (Hoenisch & Niggemeyer, 2004, p. 76)

The program uses the concept of journeying between lands to enable children to develop and construct mathematical concepts by moving from concrete to abstract mathematical forms. Children journey from the land of pictures (concrete) to the land of symbols (abstract) using activities that act as bridges to connect to core concepts of sorting, pattern, numbers, geometry, measurement, and statistics. This program is named after an activity that introduces the ‘Math King’ as a key player in a counting game: a group of children forms a circle, one child is standing in the middle as the Math King and carries a crown with the number symbols 1–6. In several turns the king counts each sixth child off. This procedure is continued until only one child is left—the new Math King.

(5) *Experiencing geometric figures* (Hoenisch & Niggemeyer, 2004, p. 89)

The program ‘Mathe Kings’ includes activities where children make geometric figures with their hands and bodies. For example, by using their teeth and their hands they deform a circular thread into a triangle. Children are also asked to lie down on the floor such that the straight bodies of three or four children form the sides of a triangle or a rectangle, or they paint geometric shapes on the floor and try to fill the interior by as many children as possible.

(6) *A tile-based game with triangles and rhombi*

Friedrich Froebel's seventh gift consists of symmetric isosceles triangles (halves of squares) and triangles with angles 90°, 60°, and 30° (halves of equilateral triangles) as tiles for making geometric shapes. This choice of tiles has turned out not to be appropriate. An anonymous author found that the halves of the regular triangles need to be replaced by rhombi with angles 45° and 135° and sides of the same length as the shorter side of the triangle. These tiles fit together very well, so that interesting figures can be made and given shapes (houses, boats, triangles, squares) filled, similar to pattern block and tangram activities.

(7) *Learning to calculate with the Mouse* (Vogel 2004)

This book subtitled 'Simple counting and calculating for kindergarten children' uses a silent problem-solving mouse character from a well-known German children's television series 'Die Sendung mit der Maus' (The Program with the Mouse). On the back cover, we read 'Learning in this way is double fun! Together with the duck and the elephant, the beloved mouse guides children through the book and motivates them for solving diverse tasks.' Each page presents tasks that are typically found in workbooks for grade 1 such assigning number symbols to sets, filling in missing numbers in sequences which follow a rule, expressing pictorial representations of addition tasks in symbols (e.g., seven balls in one row and five balls in the second row) and finding the results of addition tasks (e.g., 7 + 5 and 7 − 5).

(8) *Oware—Ghana's national game*

Oware is an ancient and widely played strategy game for two people originating in Africa that aims to provide children with basic mathematical skills and knowledge. It is played on a rectangular wooden board (Fig. 2.1). Each player has control of six small pits on one of the longer sides ('houses') and one big pit (the 'score house') on a shorter side. At the beginning, each small pit is filled with items that represent seeds, requiring 48 seeds, 24 for each player. To win the game, one player has to have more seeds than the other.

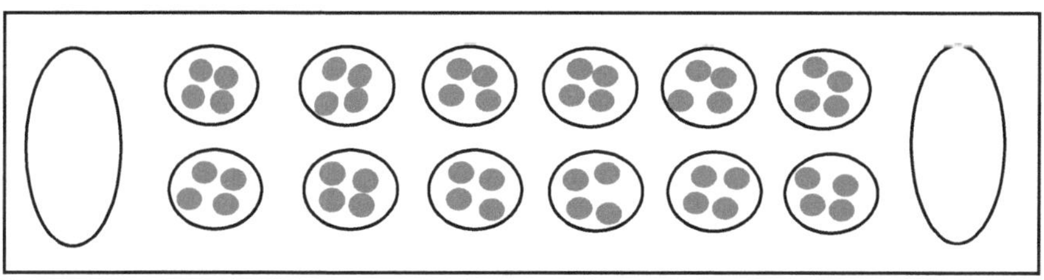

Fig. 2.1 Oware game board

Players take turns, collecting, 'sowing,' and capturing seeds. To collect, a player chooses a pit he or she controls that has one or more seeds and empties it. The seeds are then 'sown' in an anti-clockwise direction by dropping one seed in each of the pits (but not the score house) until all the seeds have been sown, leaving the original pit the seeds came from empty, as it cannot have a seed returned to it, even if there are twelve or more seeds to sow. When the last seeds have been sown, if that pit belongs to the other player and contains exactly 2 or 3 seeds, the seeds are captured by the player who has sown the seeds and are taken from the pit and placed in the score house. If the pit immediately either side of the last sown pit also contains 2 or 3 seeds and belongs to the other player, these are also captured by the sowing player. The game continues until no more seeds can be captured.

The game encourages strategies such as minimizing or maximizing seed numbers in your own pits as the number of seeds in each pit can advantage or disadvantage a player.

(9) *What the mouse is able to do* (Schmitt-Menzel & Streich, 2001)

The same mouse character used in activity (7) is also the guiding figure in this book. On one page entitled 'The mouse is planning a trip' there is a collage of pictures which is accompanied by the following lines;

'1 hammer + 4 nails + 1 rope + 1 towel = ?
6 branches + 1 sieve + 1 box = ?
1 pot + 2 cooking spoons = ?
5 cans + 1 board = ?
2 towels + 1 backpack = ?'

Children are asked to solve these equations and write the answers as numerals.

(10) *Patterns in sequences*

In 1967 Richard Feynman (1918–1988), Nobel Laureate in Physics, gave a talk to teachers in which he explained how his father had introduced him into mathematics (Feynman 1968, S. 315):

> When I was very young – the earliest story I know – when I still ate in a high chair, my father would play a game with me after dinner. He had brought a whole lot of old rectangular bathroom floor tiles from some place in Long Island City. We set them up on end, one next to the other, and I was allowed to push the end one and watch the whole thing to go down. So far so good. Next, the game improved. The tiles were different colors. I must put one white, two blues, one white, two blues, and another white and then two blues – I may want to put another blue, but it must be a white. You recognize already the usual insidiousness: first delight him with play, and then slowly inject material of educational value. My mother who is a much more feeling woman began to realize the insidiousness of his efforts and said: "Mel please, let the poor child put a blue tile if he wants to." My father said; "No, I want him to pay attention to patterns. It is the only thing I can do that is mathematics at this earliest level.

This activity could be incorporated into working with children by using red and blue counters to develop regular, linear patterns.

2.5 Evaluation of the Ten Examples

On the surface, the ten examples of mathematics experiences for young children in principle do not have significant differences. All activities deal with numbers and geometric forms although in varying ways. It is not surprising therefore that mathematically naïve users including educators, who have not considered the mathematics structures, procedures, processes, and content they engage, could put them in one pot. A closer analysis of their mathematical substance, however, reveals fundamental differences between Set 1, comprising activities (2), (4), (6), (8), and (10) on the one hand, and Set 2, comprising activities (1), (3), (5), (7), and (9) on the other hand. How those differences are construed is framed by individual perspectives about what mathematics is, which in turn reveals its accompanying structure and the relationships between its parts, and therefore how the structure and relationships can be explored, described, and communicated. Today, viewing mathematics as the 'science of patterns' is widely accepted and shared (Devlin, 1996; Hardy, 1940: Sawyer, 1955; Steen, 1998; Wittmann & Müller, 2007). From this perspective, the existence and recognition of pattern signals mathematical significance (Sawyer, 1955). Pattern in this context is conceived of as a general relationship between numbers, forms, data, functions, or more complex mathematical objects. Often the terms 'structure' or 'patterns and structures' are used synonymously. The influence of pattern and structure on how mathematics is learned and taught continues to gain traction and influence in early childhood mathematics education research and practice (e.g., the work of Mulligan & Mitchelmore, 2015, 2016), and has arguably extended the way mathematics is seen as a discipline, and seen to connect to other disciplinary thinking, including the arts (see, e.g., Sinclair, 2006).

In order to understand what mathematics is about when positioned as inherently connected to structure and pattern, it is crucial to distinguish two complementary aspects: pure and applied mathematics. Applied mathematics provides us with methods for solving practical problems in many areas. These methods are based on mathematical patterns and structures that are recognized as applicable to problem-solving. In pure mathematics, patterns and structures are researched for their own sake without any reference to possible applications. It may find a practical application, but that is not the intent. In the course of history, impressive mathematical theories with an immense richness of patterns have been created. They are by no means impractical or irrelevant by the way they were developed. On the contrary, the history of mathematics is full of instances where later on, theoretical understanding of patterns of pure mathematics have found surprising practical applications. Albert Einstein's relativity theory was based on 'Riemannian Geometry,' a branch of mathematics developed solely through the internal motivations by the mathematician Bernhard Riemann 50 years before. Now a subject of popular culture, Indian mathematician Ramanujan's pure mathematical insights continue to lead discoveries in mathematics and science (Ono & Aczel, 2016). The relationship between pure and applied mathematics is therefore as follows: Pure mathematics provides applied mathematics with a stock of patterns and structures that can be seen as possible models of real situations. The richer the stock, the better the chance of finding 'mathematical bricks' that can be used as building blocks for models for real situations. Pure mathematics can therefore be positioned as a means of forging connections to the application of mathematics to real situations, and hence providing mathematical substance that enable problem-solving.

Theories in pure mathematics have been developed in a playful way, often without the slightest reference to practical applications. From a learning perspective, 'playfulness' is an essential component for creativity and innovation and therefore necessary for adaptive solutions for the changing world we live in (Bergen, 2009). It is the capacity to 'play' with ideas that enables children to inquire into, and experiment with, possibilities. The search for patterns, their investigation, their modification, and their continuation is an inherently creative activity (Freudenthal, 1973). The danger is that the notion of play as a means of developing mathematical knowledge and understanding can be seen as only possible when anchored to 'real-world' or 'real-life' situations. Such a view narrows what can be said to characterize meaningful play and constricts the worlds that children can play into only those that mimic their lived experiences. Abstract worlds, where children can engage in playful thought about elements that are not 'real,' are bypassed as not possible, or not desirable spaces for young children to enter.

Mathematical patterns and structures must not be seen as static, narrowing concepts for young children, but as spaces that are open to active processes that can and should be explored abstractly. Our knowledge of mathematics and its application to twenty-first century problem-solving has increased exponentially in recent decades, and it is the pattern and order found in mathematics that has fueled the growth. It is the connected power of pattern and structure that can drive children's early mathematical intuitions and exploration, and develop mathematical insight and understanding (Steen, 1990). Children's active experience of the dynamic structure and pattern of mathematics provides access to the inherent aims, relationships, procedures, and processes in mathematics. Against this background, an authentic mathematical activity can be characterized by the following features:

- there are sets of 'elements' with mathematical properties, mathematical relationships, and mathematical operations.
- there are mathematical rules for operating with these element.
- the mathematical activity is always goal-directed and aims to explore patterns and structures and solve problems by using known patterns and structures.
- mathematics is hierarchically structured and structures that are already known form a basis of understanding for further development.

Similar criteria have developed from research on mathematical games played with young children. Findings support the need for game activities to be structured, and to engage mathematically correct content, for the content to be essential for ongoing mathematical learning, and for it to be connected to the actual playing of the game (Gasteiger, 2012). An analysis of the Set 1 activities using these criteria reveals that they differ from the Set 2 activities in their authenticity as mathematical activities.

If we examine Set 1 activity (2), all properties of genuine mathematics are clearly visible. When presented with an image in the page of the book, children's goals are to explore the relationships between that image and their mirror images by operating with a mirror. They discover mathematical properties and relationships between the images, for example, that the reflected images are congruent, but mirror-inverted. They use symmetry in order to reach certain objectives and experience the mathematical rules, including those for spatial positioning necessary to achieve symmetrical images. Such experiences provide excellent sensory hands on introduction to the structure and principles of bilateral symmetry which plays a fundamental role in further learning.

Repeatedly counting to six in Set 1 activity (4) is directed by a definite rule and draws children's attention and observation to the patterns behind counting off. The goal of determining the next Math King focuses children's attention on the patterns and number relationships that emerge as successive counts are performed with a reduced number of 'objects' (children). When toward the end of each round only a few children are left, the children are able to draw from the patterns and rules developed from the mathematical relationships and operations and predict who will be the next Math King. These experiences build iteration and generalization, both fundamental ideas of mathematics that children will meet repeatedly in later learning.

The Froebel inspired tile-based game in example Set 1 activity (6) introduces children to the fundamental geometric idea of 'fitting,' which covers wide parts of early geometry. Freudenthal (1971, p. 422) explains:

> In paving a floor with congruent tiles there is a leading idea, I mean fitting. It is the same as in space and it is realized as concretely. Fitting is a motor sensation. Psychologists can tell you how strongly the motor component of the personality is marked at a young age, how important motor apprehension and memory may be. Things fit. Do children ask why? Apart from a rare exception young children do not. All these miracles of our space do not seem to make any impression. But they grind as millstones. The highest pedagogical virtue is patience. One day the child will ask why, and there is no use starting systematic geometry before that day has come. Even worse: it can really do harm. Of course fitting as a leading idea requires that space is acknowledged as the home of solid bodies.

Objects in this goal-directed activity are shapes, the operation children use, composing and decomposing shapes, is again a fundamental idea of mathematics which reaches far into concurrent and future learning. Fitting the pieces together requires children to work with the relationships between various plane shapes with particular attributes, and the development of intuitive fundamental geometric spatial principles.

The game 'Oware' is an example (8) that is clearly governed by mathematical rules. Mathematical objects are the various distributions of pebbles. During the game the players are continuously challenged to determine the number of pebbles in the pits and to use this information for collecting as many pebbles as possible. Here anticipation and prediction is called for, using the development of understanding of number relationships. Continuing patterns of counters and inventing rules for producing patterns in activities (10) show all the discernible features of a genuine mathematical activity. In order to create patterns, counters of different colors have to be arranged in a sequence by following certain rules. Children have to derive a rule from the beginning of this sequence and to continue it, and they can also invent their own rules. The rules, however, provide insight into the consistency of the mathematical structure they are working with.

In marked contrast with these examples, which form the seeds of children's substantial mathematical theories, those activities found in Set 2 are mathematically limited and therefore not able to be characterized as authentic mathematical activities. Although some of them represent and engage mathematical concepts to some extent, the mathematical rules governing operations and patterns which could be explored are not visible, and there is not a worthwhile problem to be solved: there are no mathematical spaces in which children can move. Set 2 activities (1), (3), (7), and (9) in particular use popular characters from children's entertainment media to contextualize the activities to children's 'real-world' experiences. In doing so, the mathematical content is secondary to the story, which although promoting a 'playful' approach is irrelevant to any mathematical content or purpose. As a consequence, what children are expected to do is completely determined by non-mathematical frameworks and devoid of opportunities to see and experience mathematical patterns and structures. Developing understanding of mathematical elements is not the goal of these activities. In comparing Set 2 activity (5) and Set 1 activity (6), for example, the contrast is apparent. When children form polygons with their bodies in activity (5), although the goal is to explore geometric structures, there are no clearly defined vertices or angles that can be formed by human bodies. As a result, the opportunity to work with the mathematical properties and explore the relationships between vertices, angles, and area is not possible. In this context, there is no sense of a meaningful combining of geometric figures or any idea of 'fitting in a mathematical sense' as a precise motor sensation that connects to spatial relationships. Similar inattention to critical mathematical structure and concepts is also found in books for very young children where shapes are shown with rounded corners.

Activities in Set 1 differ from the activities in Set 2 also with respect to the motivation that the task may prompt in children. We argue that in the first set, an intrinsic motivation arises from the mathematical activity itself as the activities are rich and connect to mathematics as a system. Importantly, they are presented within a mathematical context. The actives begin with handling or engaging with a 'real life' experience, such as a game or picture, however, the activity goal is intentionally directed to a mathematical purpose, that is, to bring children into pure conceptual mathematical understanding. The design of the tasks unashamedly recognizes the substance of mathematics that the activity is designed to engage, and builds on this through creating spaces for children to explore, problem solve, and mentally play with pattern and structure, so that relationships can be experienced and understood.

In contrast, the activities in Set 2 are based on extrinsic motivation. Here the activities are determined by non-mathematical contexts. Even where the intention of the design is for children to learn mathematical concepts, this is achieved through contrived contexts, where mathematical concepts are characters, as in activity (3), or mathematics is used or children are instructed by characters, as in activities (1), (7), and (9). In this way, the programs from which these activities are taken are typical and exemplify 'edutainment' products which seek to educate and entertain. The sheer mass of products of this kind shows that most authors who consider themselves as competent to design early mathematics programs are not aware of the intrinsic motivations which mathematics itself offers. More importantly, it reveals that such authors are unaware as to how tasks can be designed that can bring children to abstract mathematical thinking, and that this can be achieved playfully and meaningfully. Instead they are convinced that mathematics is by its very nature a 'dry' subject that has to be motivated by stimuli coming from what adults perceive to be attractive to children. Adult beliefs that mathematics is a subject of little interest to children have long standing. Montessori was aware of it and counteracted it through the Froebel inspired materials she developed for very young children's mathematics education, which she termed 'materialized abstractions' (Montessori, 1995, p. 186).

It would, however, be wrong to completely exclude extrinsic motivation from early mathematics education. In small doses and as stimuli, it can be helpful for overcoming certain thresholds on the way to intrinsic motivation (see, e.g., the excellent neuropsychological analysis in Zull, 2002, pp. 53–55). Our argument is, however, that if mathematically founded early mathematics programs must incorporate some elements of low-level concept, such programs also require those low-level activities to be embedded in rich mathematical activities that aim to connect to mathematical pattern and structure, and be explicit in the mathematical elements and rules that children are to work with.

2.6 The Impact of Pure Mathematics on Early Mathematics Education

Early math programs that are based on an authentic view of mathematics in the sense discussed in this chapter cannot be restricted to experiences with numbers and forms found in children's daily life. In order to bring the pure aspect of mathematics to bear, it is absolutely necessary that children also experience mathematics as an 'artificial' world, that is, an abstract mathematical world that has been created by humans and that is governed by internal rules integrally connected to the pattern and structure inherent in mathematics. This position is not easily accepted in early childhood educational settings. As experiences show, early childhood educators have a clear preference for activities described as 'learning with all senses' typified by activity (5) with children forming polygons with their bodies and for 'everyday mathematics,' that is, for mathematics found in children's lived environment. Mathematically rich activities similar to the geometric tile activity (6) are often rejected as 'dry,' 'not child-centered,' and 'not motivating,' attitudes that can reflect an adult educator's own mathematical content knowledge and negative perspective on mathematics gained from their own learning experiences (Goff & Dockett, 2015). It is exactly here where mathematical enlightenment has to start. Fortunately, mathematics educators are not left alone when arguing in favor of a deeper understanding of mathematics. Significant support comes from psychology.

As early as 1978, Donaldson (1978) warned of the pitfalls of learning that focuses exclusively on children's senses and everyday experiences. She advocated for education to familiarize children with a more 'theoretical' approach to learning in their early years, and to transcend immediate experiences in order to establish some distance from the real world. Donaldson believed that it was paradoxical that our practical real-world successes, such as engineering, require working with and manipulating abstract structure and pattern found as systems, which by their very nature, require a disembodying of thinking beyond the real, sensory world we live in. If we see the interaction with abstract thought as irrelevant, inappropriate, or harmful to young children, we run the very real risk of disembodying children's thinking from the very thing that is needed—connecting pure mathematics to real-world applications. The step from daily life to 'disembedded thought' is not spontaneous, however, and Donaldson argues that is has to be supported by adults. Educators need to offer opportunities and encouragement for children to express their interest in mathematical questions they genuinely want answers to and then to mathematically motivate children by connecting them to learning tasks that are meaningful and mathematically relevant. Such connections require educators who are cognizant of the early learning mathematical knowledge and understanding we currently hold that has the power to move children mathematically beyond the 'everyday.' Children's past experiences and existing knowledge are powerful motivators to mobilize children's mathematics learning.

Donaldson's concerns and prerequisites for acquiring knowledge connect to those of Young (2010) who is at pains to distinguish knowledge from experience, and curriculum from pedagogy. Disciplinary specific concepts originate in disciplinary knowledge–producing communities, of which mathematics is one. Mathematics has specific purposes and ways for reliable generalizations to be made, generalizations that are possible because there are mathematical elements and rules underpinned by mathematical pattern and structure, whereas everyday concepts are acquired informally and ad hoc for particular problems and purposes in particular contexts, disciplinary knowledge has sets of related, organized concepts, each with distinct and explicit relationships with each other that must be taught. Understanding the concepts and their relationships therefore represents the only reliable way to extend children's learning beyond their everyday experiences. Education, Young argues, needs to support children to manage the relationship between their everyday lives and disciplinary knowledge, and to introduce children to knowledge 'which have meanings that do not derive from or relate directly to their experience' (p. 26). It is this knowledge that can forge children's identities as mathematical learners and enable children to generalize beyond their own world experiences.

Further support for this position is provided by Jean Piaget's and Bärbel Inhelder's genetic epistemology and psychology. The comprehensive studies on children's cognitive development of these ingenious researchers have convincingly shown that knowledge about numbers and forms is not derived from sensual experiences in a simple way, but is constructed in interaction with the environment. 'Penser c'est opérer' is one of the key statements of the Genevan School. In this sense, mathematical knowledge is not the result of perception, but the result of operating on mathematical objects (sets, numbers, forms) in meaningful contexts and observing what effects the operations exert on properties and relationships of the objects. Although some results and methods of the Genevan School have been questioned and in part rejected, the general idea that knowledge is not prefabricated and not just copied has remained untouched. From this 'operative' point of view Jean Piaget has argued against 'visual teaching methods' and in favor of 'active methods' (Piaget, 1972, pp. 79–81). 'Learning with all senses' is a postulate that is widely shared in early childhood education. There is no doubt that it has some value for experiencing natural phenomena. When it comes to mathematics, however, it is highly problematic. Numbers and mathematical forms are mental constructions. They cannot be perceived by the senses in principle. It is true that numbers and forms can be represented by means of certain physical materials. The core of the problem is how these materials are used, which problems are represented and tackled, which properties and relationships are relevant, and which patterns can be discovered: all of which is not determined by visual perceptions but by goal-directed mathematical activities which are embedded in mathematical structures. The earlier children become aware of the 'operative,' structured nature of mathematical concepts, the easier it will be for them to learn mathematics. This plea for pure mathematics must not be understood as unconditional. Everything depends on mathematical substance. New Math in the 1970s which was based on a hollow formal language is a striking counterexample.

2.7 Conclusion

The challenge we face in early years mathematical learning is finding a bridge between pure and applied mathematics, connections that are possible if tasks are designed using activities that have mathematical authenticity and integrity. Such things are possible if we can design activities that have at their forefront an intention to provide children with ways to develop and think with abstract models of mathematical pattern and structure, models that can be applied to the real world. The long-term success of early mathematics education stands and falls not only with the mathematical quality of curriculum and learning environments but also with the view of mathematics that early childhood educators hold. Therefore, it must be the first and foremost objective of both pre-service and in-service education to bring early childhood educators into understanding of genuine mathematics, that is, pure mathematics that can provide possible models for the application of mathematics to real situations. What is needed are courses in which educators are introduced to substantial learning environments, learning environments that embody criteria for mathematical structure and, pattern and processes as central to their design, and which are connected to and combined with reflections about the nature of mathematics. As an example of such a course, we refer to is the Mathe 2000 Early Maths program, which is described and illustrated by field experiences from several countries in Wittmann (2016).

We have learned the value of worthwhile literature in children's early learning. Leo Lionni, the well-known author of children's books, was once asked to explain what good children's literature is about. His answer: 'Good children's literature is also literature for adults' (source unknown). This statement can be transferred to early childhood mathematics education: mathematically rich and mathematically authentic early mathematics programs allow for the acquisition of a deeper understanding and a more comprehensive knowledge of mathematics, and both educator and child are richer for it.

References

Anonymous. (2006). *Wickie und die starken Männer.* Vorschulbuch. Suchen, Finden, Zählen, Zuordnen. Köln: Schwager & Steinlein.

Bergen, D. (2009). Play as the learning medium for future scientists, mathematicians and engineers. *American Journal of Play, 1*(4), 413–428.

Clarke, B., Clarke, D., Grüßing, M., & Peter-Koop, A. (2008). Mathematische Kompetenzen von Vorschulkindern. Ergebnisse eines Ländervergleichs zwischen Australien und Deutschland. *Journal für Mathematik-Didaktik, 29,* 259–286.

Department of Education, Employment and Workplace Relations (DEEWR). (2009). *Belonging, being and becoming: The early years learning for Australia.* Canberra: Commonwealth of Australia.

Devlin, K. (1996). *Mathematics: The science of patterns.* Holt Paperbacks.

Donaldson, M. (1978). *Children's minds.* London: Fontana.

Feynman, R. Ph. (1968). What is science? *The Physics Teacher, 7,* 313–320.

Freudenthal, H. (1971). Geometry between the devil and the deep sea. *Educational Studies in Mathematics, 3*(3/4), 413–435.

Freudenthal, H. (1973). *Mathematics as an educational task.* Dordrecht: Reidel.

Froebel, F. (2005). *The education of man* (W. H. Hailmann, Trans.). New York: Dover Publications (Original work published 1898).

Gasteiger, H. (2012). Fostering early mathematical competencies in natural learning situations. Foundations and challenges of a competence-oriented concept of mathematics education in kindergarten. *Journal of Mathematics Didakt, 33,* 181–201.

Gasteiger, H., Obersteiner, A., & Reiss, K. (2015). Formal and informal learning environments: Using games to support early numeracy. In J. Torbeyns et al. (Eds.), *Describing and studying domain-specific serious games* (pp. 231–250). Switzerland: Springer International.

Goff, W., & Dockett, S. (2015). Partnerships that support children's mathematics during transition to school: Perceptions, barriers and opportunities. In B. Perry, A. MacDonald, & A. Gervasoni (Eds.), *Mathematics and transition to school: An international perspective.* Singapore: Springer.

Hardy, G. H. (1940). *A mathematician's apology.* Cambridge: Cambridge University Press.

Hengartner, E., & Röthlisberger H. (1995). Rechenfähigkeit von Schulanfängern. In Brügelmann, H. u.a. (Hrsg.): *Am Rande der Schrift. Jahrbuch 6* (pp. 66–86). Bottighofen: Libelle Verlag.

Hoenisch, N., & Niggemeyer, E. (2004). *Mathe-Kings. Junge Kinder fassen Mathematik an.* Berlin: Weimer.

Krajewski, K. (2008). Prävention der Rechenschwäche. In W. Schneider & M. Hasselhorn (Hrsg.), *Handbuch der Psychologie, Bd. Pädagogische Psychologie* (pp. 360–370). Göttingen: Hogrefe.

Krieg, S., Curtis, D., Hall, L., & Westenberg, L. (2015). Access, quality and equity in early childhood education and care: A South Australian study. *Australian Journal of Education, 59* (2), 119–132.

McDonald, A., Goff, W., Dockett, S., & Perry, B. (2016). Mathematics education in the early years. In K. Makar, S. Dole, J. Visnovska, M. Goos, A. Bennison, & K. Fry (Eds.), *Research in mathematics education in Australasia 2012–2015* (pp. 165–186). Singapore: Springer.

Montessori, M. (1995). *The absorbent mind* (C. Claremont, Trans.). New York: Henry Holt and Co. (Original work published 1949).

Moss, J., Brice, C. D., & Bobis, J. (2016). Young children's access to powerful mathematics ideas. In L. English & D. Kirshner (Eds.), *Handbook of international research in mathematics education* (3rd ed., pp. 153–190). New York: Taylor & Francis.

Mulligan, J. (2015). Looking within and beyond the geometry curriculum: Connecting spatial reasoning to mathematics learning. *ZDM Mathematics Education, 47,* 511–517.

Mulligan, M., & Mitchelmore, M. (2016). *PASMAP: Pattern and structure mathematics awareness program.* Camberwell, Vic: ACER.

Mulligan, J., Mitchelmore, M., English, L., & Crevenston, N. (2013). Reconceptualizing early mathematics learning: The fundamental role of pattern and structure. In L. D. English & N. Crevensten (Eds.), *Reconceptualizing early mathematics learning: The fundamental role of pattern and structure* (pp. 47–66). Netherlands: Springer.

Ono, K., & Aczel, A. D. (2016). *My search for Ramanujan: How I learned to count.* New York: Springer.

Pauen, S. (2009). Die Wirksamkeit mathematischer Frühförderung in der Kindertageseinrichtung. In S. Pauen & V. Herber (Eds.), *Vom Kleinsein zum Einstein* (pp. 50–58). Berlin: Cornelsen.

Perry, B., & Dockett, S. (2008). Young children's access to powerful mathematical ideas. In L. D. English (Ed.), *Handbook of international research in mathematics education* (2nd ed., pp. 75–108). New York: Routledge.

Piaget, J. (1972). *Theorien und Methoden der modernen Erziehung.* Wien: Molden.

Preiß, G. (2004). *Geschichten aus dem Zahlenland.* Kirchzarten: Verlag Zahlenland.

Sarama, J., & Clements, D. H. (2009). *Early childhood mathematics education research: Learning trajectories for young children.* New York: Routledge.

Sawyer, W. W. (1955). *A prelude to mathematics.* London: Penguin.

Schmidt, S., & Weiser, W. (1982). Zählen und Zahlverständnis von Schulanfängern. *Journal für Mathematik-Didaktik, 3,* 227–263.

Schmitt-Menzel, I., & Streich. (2001). *Was die Maus alles kann.* Ravensburg: Maier.

Sinclair, N. (2006). *Mathematics and beauty: Aesthetic approaches to teaching children.* New York: Teachers College Press.

Spiegel, H. (1997). *Spiegeln mit dem Spiegel.* Stuttgart: Klett.

Steen, L. (1998). The science of patterns. *Science, 240*(452), 611–616.

Steen, L. A. (1990). Pattern. In L. A. Steen (Ed.), *On the shoulders of giants: New approaches to numeracy* (pp. 1–12). Washington DC: Mathematical Sciences Board.

Sylva, K., Melhuish, E., Sammons, P., Siraj-Blatchford, I., & Taggart, B. (2004). *The effective provision of pre-school education (EPPE) project.* London: Institute of Education: University of London.

Van den Heuvel-Panhuizen, M. (1995). *Assessment and realistic mathematics education.* Utrecht: CD-β.

Vogel, H. (2004). *Rechnen lernen mit der Maus 1.* Braunschweig: Westermann.

Walter, M. (1972). *Make a bigger puddle, make a smaller worm.* London: Evans.

Walter, M. (1984). *The magic mirror book.* London: Scholastic.
Walter, M. (1985). *The mirror puzzle book.* London: Tarquin.
Wittmann, E. Ch. (1998). Mathematics education as a 'design science'. In A. Sierpinska & J. Kilpatrick (Eds.), *Mathematics education as a research domain: A search for identity* (pp. 87–103). Netherlands: Springer.
Wittmann, E. Ch. (2004). Assessing preschoolers' geometric knowledge. In B. Clarke et al. (Eds.), *International Perspectives on learning and teaching mathematics* (pp. 113–126). Göteborg: NCM.
Wittmann, E. Ch. (2005). *Mathematics as the science of patterns: A guide for developing mathematics education from early childhood to adulthood.* Plenary Lecture presented at the International Colloquium "Mathematical Learning from Early Childhood to Adulthood" organized by the Centre de Recherche sur l'Enseignement des Mathematiques in collaboration with the Institut de mathématique de l'Université de Mons-Hainaut, Mons/Belgium, July 7–9, 2005.
Wittmann, E. Ch. (Ed.). (2016). *Kinder spielerisch fördern - mit echter Mathematik.* Seelze: Kallmeyer-Klett.
Wittmann, E. Ch., & Müller, G. N. (2007). Muster und Strukturen als Grundkonzept. In Walther, G. u.a. (Hrsg.), *Bildungsstandards Mathematik für die Grundschule: Mathematik konkret* (pp. 42–65). Berlin: Cornelsen Scriptor.
Young, M. (2010a). The future of education in a knowledge society: The radical case for a subject-based curriculum. *Journal of the Pacific Circle Consortium for Education, 22*(1), 21–32.
Young, M. (2010b). Why educators must differentiate knowledge from experience. *Journal of the Pacific Circle Consortium for Education, 22*(1), 9–20.
Young, M., & Muller, J. (2010). Three educational scenarios for the future: Lessons from the sociology of knowledge. *European Journal of Education, 45*(1), 11–27.
Young, M., & Muller, J. (2013). On the powers of powerful knowledge. *Review of Education, 1* (3), 229–250.
Young, M., & Muller, J. (2016). *Curriculum and specialization of knowledge: Studies in the sociology of education.* London: Routledge.
Zull, J. E. (2002). *The art of changing the brain.* Sterling, VA: Stylus.

Authors Biography

Dr. Virginia Kinnear is a lecturer in early childhood education at Flinders University in South Australia. She worked as a solicitor for 10 years before coming to early childhood education through studying and teaching Montessori education (3–6 years) in the USA. She then qualified and taught in early childhood teaching qualification in Australia and has 18 years experience in early childhood and tertiary teaching. Virginia's Ph.D. focused on statistical reasoning in early childhood. Her research and teaching interests are in all aspects of mathematics learning. She is particularly interested in young children's statistical learning and thinking, the development of critical and ethical thinking, and the intellectual attributes and processes that support mathematical engagement.

Erich Ch. Wittmann studied mathematics and physics at the University of Erlangen/Bavaria and finished with master degrees in both subjects. In 1967, he received his Ph.D. in mathematics (group theory) from the University of Erlangen and moved to mathematics education following Hans Freudenthal's advice. In 1969, he was appointed full professor of mathematics education at the University of Dortmund and worked in this position until his retirement in 2004. Erich Wittmann's research is based on a view of mathematics education as a 'design science.' This approach has been put to practice by him in the project 'mathe 2000' founded in 1987. The innovative textbook DAS ZAHLENBUCH originating from this project has been adapted to several countries. In 1998, Erich Wittmann received the honorary doctorate from the University of Kiel, and in 2013, he was awarded the Johannes Kühnel Prize. His international reputation has been acknowledged by the invitation to present a plenary lecture at ICME 9, Tokyo 2000.

Chapter 3
Powerful Frameworks for Conceptual Understanding

Camilla Björklund

Abstract This chapter centres around teaching and learning of mathematics in the early years and in particular the frameworks for facilitating concept development. Contemporary research conceptualizes mathematics education as orchestrating for conceptual exploration and meaning making of everyday notions. The notion chosen to work as an empirical example to illustrate recent insights from research is the notion "half"—a complex notion in its own simplicity. Different frameworks orchestrated by one teacher through five different activities aiming at the same learning object, constitute three critical dimensions in the teaching act: (1) physical appearance of items, (2) the structure of the teaching activity, and (3) the child's attention and intention. These dimensions will be described and analyzed in relation to the learning opportunities they provide.

Keywords Conceptual understanding · Half · Attention · Preschool Structure · Learning intention · Variation theory of learning · Preschool

3.1 Introduction

This chapter directs attention to the teaching and learning of mathematics in preschool and more specifically, how to understand the pedagogical framework that teachers construct in their pedagogical practice. The chapter presents a study conducted within a research project, aimed at understanding concept development through play-based activities, in order to deepen our understanding of teaching in both designed and spontaneous situations. Children's perspectives and initiatives, physical framing such as props and manipulatives and how to orchestrate learning are key points in the discussion. I will argue for the importance of a flexible and perceptive approach in mathematics education and in particular show with cogent

C. Björklund (✉)
University of Gothenburg, Gothenburg, Sweden
e-mail: camilla.bjorklund@ped.gu.se

V. Kinnear et al. (eds.), *Forging Connections in Early Mathematics Teaching and Learning*, Early Mathematics Learning and Development,
https://doi.org/10.1007/978-981-10-7153-9_3

empirical examples how sensitivity to the physical features, pedagogical structure, and children's perspectives are relevant for potential learning outcomes.

The departure point for this chapter is from the extensive body of research and theories on mathematics learning in the early years. This is followed by a focus on the child's perspective in learning about children's development and teaching for conceptual understanding that has been central for several studies conducted in the Nordic countries during the last decade. These studies provide important insights to the field of early childhood didactics, in particular the relevance of careful framing of learning situations, which will be exemplified by authentic observations from a recent research project.

3.2 Mathematics in the Early Years

Children's mathematical development is probably one of the most thoroughly researched areas in developmental and educational psychology. Many studies have their origins in Jean Piaget's extensive work on children's development of logical reasoning and arithmetical skills, confirming and extending his work (see for example Kamii, 1996). The Piagetian standpoint, in general, has been challenged not least by Donaldson (1978; see also McGarrigle & Donaldson, 1974), who argues that children show competences and cognitive abilities far more advanced than those shown in traditional experimental procedures. The characteristics of the experimenter's behavior and actions toward the task material seem to influence children more than linguistic attributes. This is probably one of the reasons why children often show very different skills in naturalistic and experimental studies. What features a child directs his or her attention to and how an instructor acts upon objects in a task and thereby calling attention to specific aspects—which is not always intended—is an issue that has to be addressed in research on young children's conceptual development (Hundeide, 1985). This issue does, however, call for attention in any teaching act with young children.

What is possible to experience and discern as important teaching aspects is also the topic of discussion in this chapter, but within a recently developed theoretical framework—Variation Theory of Learning (Marton, 2015). The theory provides a framework for explaining what is possible to learn in any particular situation, considering learners' different experiences that constitute aspects of a learning object their attention is directed toward. From this perspective learning is about becoming more aware of differentiated aspects of the surrounding world, and teaching is about providing experiences to the learner that will extend his or her way of "seeing" certain phenomena. The children's perspectives and what is made possible to discern in a specific situation are highly integrated within the framework in order to interpret their mathematical skills and knowledge. Vygotsky's (1987, 1998) discussion on children's concept knowledge and development is in many ways in line with this theory, whereby children are not seen to lack logical reasoning skills, but rather to think qualitatively different than adults do (see also

Marton, 2015; Marton & Booth, 1997). As will be shown in the empirical examples given later in the chapter, children may prefer to divide sets of half on different grounds than adults do, in ways that are equally logical from the different perspectives. These differences, however, may have consequences for what arguments and proofs are considered relevant in a problem-solving situation.

Different theoretical paradigms put light on different perspectives that are seemingly opposite to each other, with Piaget and Vygotsky as two prominent names that are often used to illustrate opposite perspectives. On a closer look, however, the theoretical frameworks found within different paradigms of research serve to focus attention on different questions of interest, without necessarily neglecting the complexity of development and learning. Instead, in using the systematic processes of science, researchers are obliged to focus the lens on demarcated research objectives. Naturally, these research questions and objectives are informed by ontological and epistemological standpoints that are made visible through the questions asked and the interpretations made of observations of empirical phenomena. The following section is therefore an overview of the most essential knowledge within the field of early mathematics learning and development, providing an eclectic picture of the state of the art.

Developmental psychology and research on cognition have provided a great number of studies that enhance our understanding of young children's development of arithmetical skills. Studies have, for example, shown that infants seem to have intuitive knowledge of changes in numerical relationships (Wynn, 1998), abilities to intuitively add and subtract large quantities (McCrink & Wynn, 2004), recognize patterns (Mandler & Shebo, 1982) and to numerically discriminate large numerosities (Lipton & Spelke, 2003), among other competences. Most of these are considered intuitive, but nevertheless provide a basis for the development and learning of more advanced mathematical principles and cultural tools such as symbols and bases.

In addition to the intuitive or even innate competences, another important contribution to the understanding of arithmetical development is the attention to number that is related to the part-whole relationships within quantities. Studies by Hannula, Räsänen and Lehtinen (2007) show, for example, that children's subitizing skills and spontaneous attention to number play a central role for their arithmetical development. Attention to number and in particular part-whole relationships within number is part of the foundation for a conceptual understanding of number, which in turn is a prerequisite for arithmetical skills (Baroody & Tiilikainen, 2003). Markman's (1979) study on children's strategies for arithmetical reasoning gives strong support for emphasizing the part-whole relationship. Children who are given the opportunity to perceive items as a collection express more powerful reasoning skills in comparison to tasks where items in a set are unconnected. Comprehension of this relationship seems to be based on the child's ability to both differentiate and integrate units into wholes. This, in turn, means that units are perceived as distinct and also as parts of an even larger whole (Hunting, 2003). How these skills develop, from the recognition of changes in quantities (adding or subtracting an item will change the original quantity in some way), to manipulation with abstracted quantities (using symbolic representations for items grouped together forming a set of objects) is a matter of debate. Some claim that development occurs due to the child acquiring the verbal number sequence (Fuson, 1988). Others argue that there are intuitive principles that emerge in early childhood while children encounter numerical phenomena in their surrounding world (Gelman & Gallistel, 1978). Yet, there are also suggestions that the conceptual development of number and arithmetical skills are strongly related to experiences offered to children where there is an emphasis on the part-whole relationship as a result of anticipated learning trajectories (Clements & Sarama, 2014).

The above-mentioned empirical studies and theoretical standpoints bear meaning to our contemporary understanding of young children developing mathematical knowledge and skills. Some intuitive processes are probably necessary in order to develop more advanced skills (Butterworth, 1999). The direction this development moves to is, however, related to the child's social, cultural, and historical context and associated expectations, values, norms, and cultural tools. Education, and not least early childhood education, is accepted as a key factor for later school performance in contemporary research (see for example longitudinal studies by Aunola, Leskinen, Lerkkanen, & Nurmi, 2004; Aunio & Niemivirta, 2010; Jordan, Kaplan, Ramineni, & Locuniak, 2009). Further, young children are often engaged in preschool or kindergarten, where they are part of a context where pedagogical practices for learning and development are at the center of daily practice.

In the Nordic countries, for example, Sweden, as many as 93% of all five-year-olds are enrolled in preschools that follow a national curriculum with guidelines for learning goals (Swedish National Agency for Education, 2011). Preschool practice is communicative and connective by nature, where children are exploring both familiar and novel phenomena together in thematic work. Communicative skills are in this respect essential to develop, alongside concept development that extends the meaning and understanding of common encountered phenomena (Pramling Samuelsson & Pramling, 2013). Many concepts that children encounter in these kinds of practices are mathematical in their nature. Children, for example, describe relationships in space, shape, or time and make comparisons between quantities and changes in quantities, which are possible to represent and communicate. Children are said to "mathematize" (see Freudenthal, 1968; Van Oers, 2013), in that they make use of mathematics as something necessary and useful for their meaning making, play, and communication. Van Oers (2013) emphasizes the social and cultural perspective on mathematics learning and suggests that early mathematics education should focus on developing children's skills to communicate about number, quantity, and space. This is accomplished when a situation requires the children to translate experiences using mathematical language and objects. This translation helps the children to interpret situations from a mathematical point of view, knowing the mathematical orientation in a certain situation and which strategies are appropriate to use.

Based on the extensive research on arithmetical development and considering the social context that most children take part of in their early years, mathematically, we have a quite complex phenomenon at our hands. Research shows that it takes many years of frequent encounters to acquire a solid understanding of a concept and the multiple ways to contextualize it, for example, number words (Fuson, 1992). Vygotsky (1987, 1998) described the development of concept knowledge as a process that departs from complex chains of words toward more logical conceptual relationships, which means that children and adults may use the same words, but have different meaning based on different logics into the concept structure. One example of such pseudo-concepts was shown by Björklund and Pramling (2014) in a study of teaching patterns. The study revealed how teachers' intentions to highlight the repetitive aspect of patterns were constrained by the children's strong attachment to the physical appearances of the objects used as representations. This calls for the need for educators to take the child's perspective and study mathematical development and learning from a pedagogical perspective, where both the content and the child are present and expected mathematical knowledge is central.

3.3 Researching Mathematics Education—Taking the Child's Perspective

A child-centered approach where children's perspectives were of central interest was conducted in an explorative study published in 2007 (Björklund, 2007, see also Björklund, 2008). This research explored the mathematics that toddlers in two Finnish preschools encountered. The mathematics was defined by adults but took an ontological approach that enables children's mathematizing or sense making of mathematical relationships to have equal value to an adult-based interpretation. This study revealed multiple occurrences where mathematical concepts were used by the teachers in everyday conversations with or about the children. Many of these concepts were then used and made meaning of by the children in their interaction and play with other children, or in dialogues with adults. The most prominent result of the study is perhaps that children and adults often use similar words and utterances, that are necessary for communicating about common phenomena, but when challenged, children as young as two or three years old showed both determination and also the skills to express their way of understanding. One example is given in Björklund (2007, my translation):

> Elisa (2 years 7 months) puts down three toy figures on a table and says: "Here are two kitties". Adam (3 years 1 month) says: "No, it's not like that, it's this many" showing three fingers. Adam then says: "One, two, three" at the same time he raises one finger at a time once more.

The example above shows toddlers responding to each other about mathematical content, in particular their motivation to express their understanding to others when differences in meaning occur. Further, the children were determined to express themselves in ways that communicated mathematical meaning and that were aimed at convincing another child as to the accuracy of their reasoning.

The study in Björklund (2007) also revealed examples where the teacher's responsiveness to the child's perspective and expressed meaning was crucial for dialogues to proceed and for a common (or more developed) understanding to be made possible. This insight from the many empirical examples in the study was followed up in several studies conducted in close collaboration with preschool teachers and their practical everyday work with mathematics education in early childhood. One aim was to further develop teachers' professional work with young children and mathematics but also to more thoroughly analyze the teaching-learning process. Early childhood education is communicative in nature and many concepts are used on a daily basis, mostly without reflection on how children (or colleagues!) may have a different understanding of the same idea. A battery of studies has therefore been conducted to enhance the knowledge among preschool teachers about the conceptual complexity in mathematical meaning. One good example is the concept of size, where notions such as "large" and "small" are considered by many teachers to be both familiar to young children and easy to understand. However, as shown in Björklund (2012), without the introduction of a concrete reference, the relative meaning of the concept generated difficulties for toddlers. A large object is only large compared to a smaller object, which has to be identified by both participants in a dialogue. Further, the study problematized the representative objects that can create conceptual challenges and, to different extents, structure and emphasize the concept as content for learning (Björklund, 2014a, 2014b). The structure that the physical context provides, including representative objects, is thereby important to consider in mathematics teaching in the early years.

Even when teachers take great care in choosing representative objects in mathematics, there are challenges in how to teach in early childhood education. How the content is presented to the children and what leaning is made possible as a result are crucial questions for any pedagogical action or study. In a recent project, this was illustrated through authentically designed activities in a Swedish preschool. A theoretical framework (Variation Theory of Learning, see Marton, 2015) was implemented by preschool teachers and the process of concept development among children was studied. The results from the project reveal that mathematical concepts, due to their complexity, need structure and a framework that emphasizes critical aspects of the concept. In this way, children are provided with opportunities to discern what they have not previously been aware of. The project evidences the necessity for teachers to be aware of the child's perspective and current knowledge (how the child understands a concept), and also how different activities, and not least how differently designed activities, can frame and challenge the child's conceptions in different ways, and with different learning outcomes (see Björklund, 2015; Björklund & Pramling, 2014). These studies, among others, highlight that the framework of pedagogical activity is multifaceted and demands attention to ways of facilitating learning and development. As discussed above, the physical appearances of the objects are one dimension, the way an activity is structured and thereby provides different aspects to be in the center of exploration is another. A third dimension consists of the child's directed attention, and not least, the child's intentions in an activity.

In the following empirical example, borrowed from the project *Learning about Space,*[1] we will follow one teacher's planned and pedagogical work with the concept of "half." The learning situation was carefully designed to enable analysis of the teaching and learning process. Results from the study can inform early childhood education practice that includes both planned and spontaneous learning situations. The discussion centers on the above-mentioned critical dimensions of the pedagogical framework that has been shown in earlier studies and published work to provide meaning for the learning outcomes, in addition to the theoretical framework that is intended to be implemented by the teachers. The empirical example is chosen because analyses of the framework dimensions emerge in the activities and are easily recognized. To make a comprehensive overview, the activities are framed by: the items that are used, their purpose for structure and relevance to maintain dialogue and common focus of the chosen concept, and also the learning opportunities the framework (constituted by the items and the relevance structure) provide. This overview is supplemented with excerpts from the dialogues when the children and teacher explored and reasoned about the concept within the activities.

3.4 Empirical Example of Frameworks for Orchestrating Learning

The learning situation in the following example consists of five activities where the teacher worked with four children, two girls and two boys, aged 4–5 years old. The learning goal is the notion *half.* We will follow all five activities through, focusing on which items are used and for what purposes. This will further lead to conclusions of how the learning object, or critical aspects of the same, are made discernable by the children. In accordance with the Variation Theory of Learning, the last conclusion is essential: teachers' intentions do not always meet children's directed attention (Marton & Tsui, 2004). Attention directed at a shared learning object is thereby a key feature of the learning process.

In accordance with the theoretical framework adopted in the project, the teacher offers the children opportunities to discern what is critical for the meaning and use of the specific concept "half." This was achieved through her choice of manipulatives but also through questions and suggestions that emphasized critical mathematical meaning to the children. In the five activities, the teacher offers the children different problems to solve that promote verbal explanations. These explanations reveal the children's understanding of half and also provide the children with an opportunity to communicate their meaning in ways that make their thoughts explicit. This is a delicate way of orchestrating learning, but one that is necessary for conceptual development.

[1]Funded by the National Research Council in Sweden, grant nr. 724-2011-751.

Objects as *props* are an important feature of the teaching acts in preschool, since their role is both to frame the learning situation, providing a context of relevant meaning, and to possibly make some aspect of the learning object explicit. In the teacher's activities, we can see a variety of props and manipulatives. These may at a closer analysis provide children with different learning opportunities. The framework for conceptual development includes three dimensions as follows: first, the *props* and what physical features they offer to experience and exploration; second, the *relevance structure* of the activity that gives meaning to the problem to solve; and third, the child's *directed attention* and initiatives that is decisive for what the child experiences and how these experiences are extended.

3.4.1 Activity 1

This first activity is a task where two lambs have one grass straw to share (see Table 3.1). Both lambs are hungry and want equal amount of the straw.

Preschool children are familiar with this kind of task, since they often engage in dividing acts where they prefer fair shares. The task is demanding due to the challenge of dividing in half without the use of discrete number or measurement tools. This challenge, however, also offers the opportunity to discuss and explore the meaning of the notion "half."

The choice of props, two lambs and one straw, directs attention to equality in the meaning of half. The children are encouraged to try dividing the grass straw by cutting it into two parts and comparing the lengths of each part. The children quickly realize that the cut made by one child resulted in two non-equal straws, which in turn resulted in a discussion about how to make the two straws equally long, and when cutting off a smaller part, what should they do with the leftover part.

> Julia (teacher): "Here are two lambs" (placing two toy lambs on the floor in front of the children). "They have a long straw" (picks up a green pipe cleaner from a box). "But they both want to eat. They are really hungry and want half of the straw each."
> Linus: "You have to cut it!"
> Elin picks up the straw, holds it between her long and index finger: "Maybe cut like this?"
> Linus cuts the straw with scissors approximately in the middle.
> Julia: "Did they get half each now?"
> Linus: "Yes, NO! This one is a bit longer" (picks up the longer part of the two cut pieces)

Table 3.1 Items and directed attention in the task where two lambs share a grass straw

Items	Relevance and meaning	Directing attention to the learning object
Two lambs	The lambs are hungry and need help to divide food	Two subjects, emphasizing the idea of sharing a quantity equally between them
One grass straw	The object to act upon in order to solve the problem and help the lambs	When dividing the straw, the idea of equal lengths in contrast to equally many parts is discerned

The children figure out that they can cut the pieces equally long by measuring them next to each other. This results in two equally long pieces and an additional short piece. Hugo leans over the short piece and cuts it in two, giving each lamb one short piece in addition to the long ones they had already.
Linus: "There, now they have half!"
Julia: "Now they have half each, how do we know that?"
Linus: "Because those are equally long" (pointing at the long piece in front of each lamb) "and those are equally long" (pointing at the short piece in front of each lamb).
Malva: "Then it became four such small ones" (pointing at the pieces in front of the lambs).

The excerpt shows examples of the conversation about the problem of sharing a grass straw equally between to two hungry lambs. The children's attention is directed toward cutting the pieces, struggling with what to do with the cutoff pieces, since the task was to use all of the straw. A critical aspect of the task was the focus on either lengths or number of parts. As the excerpt shows, children can focus on different aspects and reason differently due to their directed attention. When discussing the meaning of half, it was necessary that both the teacher and the children could find and place the same aspect in the foreground, and to also contrast length and number of parts in order to make them discernable to the children. Some children discerned the number of parts before their lengths and others focused on the lengths only, but both aspects were necessary to explore and develop a deeper understanding of the notion. The grass straw to be shared by two hungry lambs provided a relevant structure for the children to explore the concept of half in ways that facilitated conceptual development.

3.5 Activity 2

The second activity (see Table 3.2) adds another dimension to the meaning of half, which is emphasized through the teacher's use of different props and only one character this time. The character is Snick the squirrel. The children are all familiar with the fact that squirrels collect food for winter. The idea is to divide the collected food in two parts, and it enacts a natural event in the narrative told to them. The choice to have different kind of items to eat, berries and acorns, enables another important aspect of the notion half to emerge for exploration: the items may be divided in half in relation to their size or to their number.

The teacher's learning intention was to focus on the numerical dimension, but the children raised size as critical in making the subsets. The two acorns were similar in appearance, so the abstraction of the physical features was even more evident, as the contrast available in the choice of items made it possible to discern the variation in size and color, and to further explore the impact the division may have. This is a quite complex task, since some features do have relevance for the meaning of half but some (such as color) do not. The teacher encouraged the children to find out different ways of creating subsets that would emphasize the abstract aspect of half. However, not all children accepted that items can constitute

Table 3.2 Items and directed attention in the task where Snick the squirrel collects food

Items	Relevance and meaning	Directing attention to the learning object
Snick the squirrel	A hungry squirrel who saves some food for tomorrow	Squirrels need to eat but also save some for later, a reason for dividing the collected food into half
One blue pearl One red pearl Two acorns	Squirrels collect berries and nuts	The division may be done in different ways, emphasizing abstraction of the item's physical features.

subsets composed in different ways, which was important for the teacher to observe and highlight in their dialogue.

Julia (teacher): "Let's pretend these are berries. A red berry and a blue berry. And then Snick has two acorns. But Snick can only manage to eat half of his food."
Linus: "Then he can only eat two."
Elin: "I think he should eat one wine berry, one of those and an acorn" (holds up an acorn and a blue pearl).
Julia: "Mm, a blueberry and an acorn. Is that half? How do you know it is half?"
Elin: "He could take half of an acorn and half of the other."
Julia: "Can you divide in any other way, still making it half? Can you do like this, if I think Snick should eat two berries?" (picking out the two pearls)
Elin: "Nooo."
Julia: "Has Snick eaten half of his food then?"
Hugo and Malva: "Yes."
Julia: "How do you know?"
Linus: "But they are so small and those are large!" (pointing first at the pearls and then at the acorns)
Julia: "Malva, why do you think Snick has eaten half then?"
Malva: "Because it's two."
Linus: "I said, those are small" (pointing at the pearls) "and those are large" (pointing at the acorns) "he can eat two of those [acorns] because they are large."

The concept of half is multifaceted, meaning that a quantity may be divided according to spatial or numerical features, giving very different results. These are, however, important to make visible and the teacher's choice of items worked well to emphasize this critical aspect. The framework of the problem to be solved seemed to attract the children, who found it relevant to help the squirrel save half of his food for winter.

3.5.1 *Activity 3*

The props the teacher chose for the third activity were similar in size but differ in types of species. The activity is reminiscent of the previous one in that there is a quantity, a whole, which is to be divided in two equal sets (see Table 3.3). The

Table 3.3 Items and directed attention when dividing animal figures into folds

Items	Relevance and meaning	Directing attention to the learning object
Eight sticks	The sticks are ordered in two squares, illustrating folds that contain animal figures	Marks off the sets of items, which belong to which set
Two cow figures Two horse figures	Cows and horses can be found in folds, they are usually found in groups	Division into subsets emphasizing abstraction of the physical features of the items

teacher framed the task explaining that the animals live spaces that are too cramped and need more space, so a second fold needs to be created.

One child suggested that they build each animal a fold, an idea the teacher had to decline due to the limitation of eight sticks to make folds from. This is an important limitation to the learning activity, as following the child's suggestion would have lost attention to the conceptual learning goal of half, of which the division into two sets is central. The idea of making folds for the groups of animals is carefully planned. In this way, it is possible to keep the number of subsets invariant, but let the items within each subset vary. The children were encouraged to figure out the ways the animals can be grouped and still be divided so that half are in each fold. First, they responded by grouping, keeping the family resemblance with cows and horses in separate folds. A suggested change in the ways the animals can be divided resulted in the children suggesting a change of folds, as mixing horses and cows was not initially considered an option.

> Elin takes one horse in each hand and moves them from the original fold containing two cows and two horses into the new fold.
> Julia (teacher): "Now you moved the horses. That made it half, too! Now, the horses can move back, is it possible to move the animals in any other way to divide them in half, Malva what do you think?" (moving the horses back to the original fold)
> Malva shakes her head: "No."
> Julia: "Mm, but they aren't comfortable living together, the two horses. Can you move them in any other way?"
> Elin: "That horse is sad!"
> Hugo takes one cow in each hand and moves them to the new fold.
> Julia: "Now, that made it half too."
> Elin: "I know, we can make a stable for that one" (pointing at one horse) "a stable for that one" (pointing at the other horse) "that could be a stable for that one and that could be that one's" (pointing at the cows).
> Julia: "But there are only two stables."
> Linus: "But I know! I know a thing, this one stays with that one" (pointing at one horse and then at one of the cows) "and I thought that one stays and that one moves over there" (changes the places of the cow and the second horse). "That's what I thought. Because they are good friends."

Julia: "Is it still half?"
Linus: "Yes."
Elin: "'Cause they are the children" (pointing at the figures in the second fold) "and they're the adults."

The most prominent insight from this activity was that the features of the objects, the animals, were irrelevant for the mathematical operation. This is a crucial aspect in terms of abstraction development, and we can see in the episode children struggling to find the resemblance in features, even though they are capable of following logical reasoning that four animals divided into two equal sets will contain two animals. The physical appearances set up challenges but are also enabled concept development, since the critical aspect of abstraction was made visible and could therefore be explored and meaning made of it. Further, we can see that the choice of items was important for determining the direction children's reasoning can take, for example, that group inclusion can be interpreted differently, but can still work as an argument for the children's own reasoning.

3.5.2 *Activity 4*

In the fourth activity, the teacher presented a figure of Pippi Longstocking. The children are all familiar with this character who in the stories is the strongest girl in the world. She is carrying a bag of gold coins but in the activity she only wants to carry half of them, a problem the children are invited to help her solve (see Table 3.4).

The children quickly counted six coins that Pippi brought with her and divided them into two sets of three coins each. The coins all looked the same, so there was no direct focus on the possible impact of the manipulatives' features such as color or size. The numerical dimension, in this sense, is in the foreground. One child took the initiative to sort the coins into a pattern like the six on a dice, which seems to help the children first and foremost to discern a plausible division into the subsets of three but also helped the teacher to offer a contrast in pattern such as one row of four and one row of two coins. This supported visualizing the equality that the children

Table 3.4 Items and directed attention in a familiar story about the strongest girl in the world and her golden coins

Items	Relevance and meaning	Directing attention to the learning object
Pippi Longstocking figure	The figure is known to the children as the strongest girl in the world	Pippi wants to only carry half of her gold
Six similar golden coins.	Pippi carries a bag of golden coins with her	Six is divided into sets of three through ordering the coins in two rows

apparently had no trouble discerning. The similarity in features of the coins and the ability to arrange them in patterns supported their reasoning about the numerical meaning of half.

Julia (teacher) pours out some coins on the floor in front of the children: "Were there six coins?"
Linus: "Yes, then it's three on each."
Julia: "How did you know? I had in mind that you could help Pippi because she only manages to carry half of the money."
Hugo: "'Cause three plus three is six."
Julia: "You were very fast. How did you know that three was half then?"
Linus: "Well, they were one, two, three" (counting and pointing at one row of coins) "and one, two, three" (pointing and counting the other row).
Julia: "Why is this not half then?" (ordering the coins in rows of four and two)
Linus: "'Cause it is six, she cannot carry that much."
Julia: "But if she says that she can carry four, is that half?"
Linus: "No. Three is half."
Julia: "How do you know?"
Linus: "Because it is equally many coins."

This activity centers on the core meaning of "half." The items (coins) are the same and enable the children to focus on numerical aspects of half, and the meaning of half and not half. The latter is an important contrast to make use of in teaching, as it makes explicit what constitutes the meaning of the notion.

3.5.3 *Activity 5*

The fifth and last activity brings to the fore a dimension of the notion half similar to that found in Activity 2. The activity was framed by a narrative of two pigs who found a banana and an apple that they both want to eat and so they have to share the fruit by dividing it into half (see Table 3.5).

The banana and apple were cut out in colored paper, and the children were asked to cut the paper fruit in half.

Table 3.5 Items and directed attention when dividing fruit for two pigs

Items	Relevance and meaning	Directing attention to the learning object
Two pigs	Two pigs want to share fruits	To share equally, each subject should have half of the items
One paper banana One paper apple	The children often participate in dividing fruit at mealtime	Dividing two-dimensional items into half emphasises area in contrast to discrete numbers

Malva cuts the paper apple in two pieces and places one piece in front of each pig. Hugo puts the banana pieces in front of the pigs.
Linus: "But that one is bigger!" (pointing at one piece of the apple)
Hugo: "Cut off a piece here" (pointing at a corner of the larger piece of the paper apple), "a little bit, there." (Hugo gets the scissors from Malva and cuts off a corner)
Elin: "What are we supposed to do with that piece?"
Julia (teacher): "Hugo, how did you think, why did you cut off that piece?"
Hugo: "To make them equal."

All of the activities were similar in their focus on equality in the subsets, but this last activity provided certain challenges in that the framework only enabled approximate measurement of the size of the pieces, as the task is about dividing a two-dimensional non-symmetric figure. The children then turned to strategies they had explored and mastered in earlier activities, where a numerical aspect has proven to be successful (Activity 4) or at least able to be reasoned about (Activity 1). How a task is framed through the choice of items and the effect on learning opportunities is therefore a highly relevant issue to take into account in early childhood mathematics education.

3.6 Conclusion and Pedagogical Implications

The examples of activities from the study about the concept of half showed different ways of working with the same idea, particularly when the framework for engaging the concept is constituted by the *items* and the *relevance structure* of the activity (usually presented as a narrative) and the *directed attention* toward the learning object are central. Mathematical concepts are often complex in their meaning and even if the concept is used in everyday communication, the meaning may very well be experienced differently by adults and children, as Vygotsky concluded (Vygotsky, 1987, 1998). When approaching everyday concepts with knowledge of their complexity, it becomes possible to make the concepts an object for learning, and an opportunity to explore how children experience and make meaning of it. Children's perspectives and initiatives are in this sense essential in a teacher's pedagogical work. Through the designed activities presented in this chapter, knowledge was gained about children's understanding of a particular concept, "half." In everyday activities and dialogues, differences in ways of understanding may pass by unnoticed. The designed activities made such differences visible. These insights are important for teachers' interaction with children in other situations where a specific concept may be used or become a focus of inquiry, initiated by the teacher or by the children. Differences in ways of understanding are according to the Variation Theory of Learning (Marton, 2015) important to explore in educational practice, since they often provide opportunities to discern different aspects that are critical to the learning experience. For example, when talking about "half," whether the spatial or the numerical aspect that the focal point is critical, and this difference can be made visible when children encounter the half in a meaningful context with props that emphasize these specific, mathematically necessary aspects.

The contextual framework is in the center of interest and as shown in the examples above, the framework works in two ways; as a relevance structure and as means to emphasize specific aspects of the learning concept. The relevance structure is important, whether it is to gain insight into the learning processes using a designed research study, or in daily educational practice, since it frames the problem or task in such a way that it makes sense to reason about it in particular ways. This framing is also important for teaching toward a goal, in this case the concept of half. Children are creative in finding novel solutions to problems, but not all solutions will support them developing the intended concept (Björklund & Pramling Samuelsson, 2013). This is shown in the ways children in the examples used earlier strategies to solve a problem when the framework does not provide opportunities for critical aspects to be experienced and explored. Further, we can see that different manipulatives direct attention to different aspects of the concept, essential for orchestrating learning in accordance with Variation Theory. These aspects are important to recognize, since they constitute what is possible to learn in a particular situation. In other words, there is a need for mathematical awareness among the teachers to recognize which aspects children already discern and which are "invisible" but perhaps necessary to bring forth with the help of specific props or relevant structure. How to orchestrate learning through careful choices of narratives and props is thereby an essential part of teaching mathematics in the early years. This is of course applicable to learning and teaching any mathematical notion, not only "half."

Several studies have commenced in recent years in the Nordic countries using these insights, based both on theoretical conjectures and earlier empirical findings. In these studies, preschool teachers are working closely with researchers to develop knowledge of children's concept development in mathematics and the professional work of the preschool teachers. A longitudinal study emphasizing number sense is also in process, which will reveal any lasting effects of an early goal-oriented theory-driven pedagogical practice, similar to the one described in this chapter. We have knowledge and empirical evidence that early experiences of mathematics have positive long-term effects, but we need more studies to reveal what aspects and frameworks facilitate learning for all children in ways that embrace the children's perspectives and lived experiences.

References

Aunio, P., & Niemivirta, M. (2010). Predicting children's mathematical performance in grade one by early numeracy. *Learning and Individual Differences, 20*(5), 427–435.

Aunola, K., Leskinen, E., Lerkkanen, M.-K., & Nurmi, J.-E. (2004). Developmental dynamics of math performance from preschool to grade 2. *Journal of Educational Psychology, 96*(4), 699–713.

Baroody, A. J., & Tiilikainen, S. H. (2003). Two perspectives on addition development. In A. J. Baroody & A. Dowker (Eds.), *The development of arithmetic concepts and skills: Constructing adaptive expertise* (pp. 75–125). Mahwah, NJ: Erlbaum.

Björklund, C. (2007). *Hållpunkter för lärande. Småbarns möten med matematik* [Conditions for learning. Toddlers encountering mathematics, in Swedish]. (Diss.) Åbo: Åbo Akademi University Press.
Björklund, C. (2008). Toddlers' opportunities to learn mathematics. *International Journal of Early Childhood, 40*(1), 81–95.
Björklund, C. (2012). What counts when working with mathematics in a toddler-group. *Early Years. An international Journal of research and development, 32*(2), 215–228.
Björklund, C. (2014a). Less is more—mathematical manipulatives in ECE. *Early Child Development and Care, 184*(3), 469–485. doi:10.1080/03004430.2013.799154.
Björklund, C. (2014b). Powerful teaching in preschool—a study of goal-oriented activities for conceptual learning. *International Journal of Early Years Education, 22*(4), 380–394. doi:10.1080/09669760.2014.988603.
Björklund, C. (2015). The trouble with double—preschoolers' perception and powerful teaching strategies. In K. Krainer, & N. Vondrová (Eds.), *CERME9 Proceedings of the Ninth Congress of the European Society for Research in Mathematics Education* (pp. 1898–1905). Prague February 4–8, 2015.
Björklund, C. & Pramling, N. (2014). Pattern discernment and pseudo-conceptual development in early childhood mathematics education. *International Journal of Early Years Education, 22*(1), 89–104. http://dx.doi.org/10.1080/09669760.2013.809657 First published online: 27 June 2013.
Björklund, C., & Pramling Samuelsson, I. (2013). Challenges of teaching mathematics within the frame of a story—a case study. *Early Child Development and Care, 183*(9), 1339–1354. doi:10.1080/03004430.2012.728593.
Butterworth, B. (1999). *The mathematical brain*. London: Macmillan.
Clements, D. H., & Sarama, J. (2014). The importance of early years. In R. E. Slavin (Ed.), *Science, technology and mathematics (STEM)*, pp. 5–9. Thousand Oaks, CA: Corwin.
Donaldson, M. (1978). *Children's minds*. Glasgow, UK: Fontana/Collins.
Freudenthal, H. (1968). Why to teach mathematics so as to be useful. *Educational Studies in Mathematics, 1*, 3–8.
Fuson, K. (1988). *Children's counting and concepts of number*. New York: Springer.
Fuson, K. (1992). Relationships between counting and cardinality from age 2 to age 8. In J. Bideaud, C. Meljac, & J-P. Fischer (Eds.), *Pathways to number: Children's developing numerical abilities* (pp. 127–150). Hillsdale, NJ.: Lawrence Erlbaum Associates.
Gelman, R., & Gallistel, C. (1978). *The child's understanding of number*. Cambridge, Mass.: Harvard University Press.
Hannula, M., Räsänen, P., & Lehtinen, E. (2007). Development of counting skills: Role of spontaneous focusing on numerosity and subitizing-based enumeration. *Mathematical Thinking & Learning, 9*(1), 51–57.
Hundeide, K. (1985). The tacit background of children's judgments. In J. V. Wertsch (Ed.), *Culture, communication, and cognition: Vygotskian perspectives* (pp. 306–322). New York: Cambridge University Press.
Hunting, R. P. (2003). Part-whole number knowledge in preschool children. *Journal of Mathematical Behavior, 22*, 217–235.
Jordan, N., Kaplan, D., Ramineni, C., & Locuniak, M. (2009). Early math matters: Kindergarten number competence and later mathematics outcomes. *Developmental Psychology, 45*(3), 850–867.
Kamii, C. (1996). Piaget's theory and the teaching of arithmetic. *Prospects, 26*(1), 99–111.
Lipton, J., & Spelke, E. (2003). Origins of number sense: Large-number discrimination in human infants. *Psychological Science, 14*(5), 396–401.
Mandler, G., & Shebo, B. (1982). Subitizing: An analysis of its component processes. *Journal of Experimental Psychology: General, 111*(1), 1–22.
Markman, E. (1979). Classes and collections: conceptual organization and numerical abilities. *Cognitive Psychology, 11*, 395–411.
Marton, F. (2015). *Necessary conditions of learning*. New York: Routledge.

Marton, F., & Booth, S. (1997). *Learning and Awareness*. Mahwah, NJ: Lawrence Erlbaum Associates.

Marton, F., & Tsui, A. B. M. (Eds.). (2004). *Classroom discourse and the space of learning*. Mahwah, NJ: Lawrence Erlbaum.

McCrink, K., & Wynn, K. (2004). Large-number addition and subtraction by 9-month-old infants. *Psychological Science, 15*(11), 776–781.

McGarrigle, J., & Donaldson, M. (1974). Conservation accidents. *Cognition, 3*(4), 341–351.

Pramling Samuelsson, I., & Pramling, N. (2013). Orchestrating and studying children's and teachers' learning: Reflections on developmental research approaches. *Education Inquiry, 4*(3), 519–536. doi:10.3402/edui.v4i3.22624.

Van Oers, B. (2013). Is it play? Towards a reconceptualisation of role play from an activity theory perspective. *European Early Childhood Education Research Journal, 21*(2), 185–198.

Vygotsky, L.S. (1987). *The collected works of L. S. Vygotsky, Volume 1: Problems of general psychology* (R. W. Rieber & A. S. Carton, Eds.; N. Minick, Trans.). New York: Plenum.

Vygotsky, L.S. (1998). *The collected works of L. S. Vygotsky, Volume 5: Child psychology* (R. W. Rieber, Ed.; M. J. Hall, Trans.). New York: Plenum.

Wynn, K. (1998). Numerical competence in infants. In C. Donlan (Ed.), *The development of mathematical skills* (pp. 3–25). Hove: Psychology Press.

Author Biography

Dr. Camilla Björklund is an Associate Professor in education working as Senior Lecturer in Young Children's Mathematics Education at the University of Gothenburg, Sweden. Her research interests include young children's encounters with mathematics in early childhood education, mathematical reasoning in the early years, teaching strategies and professional development in early childhood mathematics education, and integration of mathematics in thematic work in preschool. She has frequently published scientific articles and chapters to report research studies conducted in the Nordic countries, both small-scale studies and large-scale longitudinal studies. Her research is also published in popular scientific books and articles, influencing practice, and teacher education.

Chapter 4
Building Connections Between Children's Representations and Their Conceptual Development in Mathematics

Janette Bobis and Jennifer Way

Abstract Young children's representations of their mathematical experiences occur as a natural part of their everyday lives. They can occur in a variety of forms, including their drawings, play, writing, gesturing, and more recently, digital productions. These representations are an essential part of young children's general cognitive, social, and emotional development. During the early years of schooling, children are expected to formalize their representations through increased use of symbols, conventional structures, and mathematical language. Supporting children to make explicit connections between their representations and mathematical concepts is an essential activity for early childhood teachers. Both researchers and teachers need to understand the nature of representations and the role they play in young children's mathematics learning. In this chapter, we focus on the mathematical representations produced by children in their first two years of school (five- to six-year-olds). We revisit data from our previous research to generate new interpretations and propose new lines of inquiry that aim to extend our understanding of student-generated representations. In doing so, we also highlight the important role early-year teachers play in helping to build connections between young children's representations and their conceptual development in mathematics.

Keywords Representations · Gesture · Drawing · Conceptual understanding Early years of school

J. Bobis (✉) · J. Way
Faculty of Education and Social Work, University of Sydney, Sydney, Australia
e-mail: janette.bobis@sydney.edu.au

J. Way
e-mail: jennifer.way@sydney.edu.au

V. Kinnear et al. (eds.), *Forging Connections in Early Mathematics Teaching and Learning*, Early Mathematics Learning and Development,
https://doi.org/10.1007/978-981-10-7153-9_4

4.1 Introduction

Young children naturally create a variety of representations of their experiences and thoughts. Representations can be insufficient for growth in mathematical understanding because they often do not require children to 'mathematize' a situation (Ginsburg, Lee, & Boyd, 2008). To mathematize, children need to connect their representations to mathematical concepts in more explicit ways—a process that normally requires careful scaffolding from a more knowledgeable adult as children move from 'emergent models' toward more formal mathematical reasoning (Gravemeijer, 2009; Meyer, 2001). The importance of making the mathematics explicit in teaching young children is well established and an essential role for early childhood teachers (Moyles, Adams, & Musgrove, 2006; Papic, 2013; Papic, Mulligan, & Mitchelmore, 2011). Support is particularly important in the early years of school when children encounter more formal and conventional symbolic representations. Yet there is so much more that researchers and teachers need to know about the complex role that representations play in children's learning of mathematics—knowledge that is necessary if we are to develop pedagogies that will overcome the difficulties many children face when learning mathematics. We agree with Goldin and Shteingold (2001) that 'the apparent limitations in some children's understanding are *not intrinsic*. Rather they are the result of internal representation systems that are only partially developed, leaving long-term cognitive obstacles and associated affective obstacles' (p. 3).

Through years of experience in various education contexts and research inquiries, both authors have come to realize the fundamental importance of mathematical representations in almost every facet of mathematics education. We have noticed the pervasive use of representations, by both researchers and teachers, often without consideration of their significance in learning processes. For example, work samples are routinely collected from children for assessment, classroom texts and worksheets are filled with diagrams, and digital resources abound in multiple representations. We believe that frequent use of such representations is made without adequate consideration for either the role they play in children's conceptual development *or* how children's minds create, interpret, and process representations. With these concerns in mind, we outline some theoretical perspectives, then re-analyze and reflect on two episodes selected from our prior research to illustrate the type of insights gained. The chapter concludes by highlighting our current research directions and proposing some practical implications for early-year teachers.

4.2 Representations in Mathematics Education

Representations are an integral part of learning mathematics. The critical importance of representations lies in the fact that mathematics essentially consists of *ideas* that are neither directly visible nor tangible. To access mathematics and to work with mathematical processes, we must create representations using signs, symbols, and conventions. These representations can take a variety of forms, such as physical objects (e.g., concrete materials, models), drawings, formal diagrams (e.g., 2-D shape constructions, number lines, graphs), mental images, mathematical–symbolic notations (e.g., numerals, algebra), or various types of verbal statements. While there may be general agreement that representations are used during mathematics instruction because they play an important role in the development of students' conceptual understanding (Braswell, 2015), the actual pedagogical approach for utilizing representations has been debated for many years (Chao, Stigler, & Woodward, 2000). For example, the best approach for using concrete and semi-concrete (structured) materials for concept development is still disputed. Some advocate repeated use of the same structured representation, while others believe multiple representations of the same mathematical idea are required. The reality may be that both approaches are needed to support different concepts and cognitive processes (Chao et al., 2000).

Whichever view is held, teachers of mathematics use external representations, such as ten frames, number lines, and symbolic notations, to exemplify concepts and processes, providing a visual image that is believed to assist students to 'see' underlying structures and highlight important mathematical relationships that can be internalized by the learner (Obersteiner, Reiss, Ufer, Luwel, & Verschaffel, 2014). It is important to realize that external representations in themselves do not possess or convey the mathematical concepts or processes they are meant to illustrate. This means there is no guarantee that students will grasp the underlying mathematics just because an external representation is employed during instruction. In fact, some research finds no learning benefits in the use of external representations (e.g., Ginsburg et al., 2008; McNeil & Jarvin, 2009). Often the fault may not be in the representations themselves, but rather in the assumption that students can make sense of the representations and 'see through' the representation to comprehend the mathematical ideas (Duval, 1999; Steenpass & Steinbring, 2014). Representational competence, however, goes beyond simply learning about and being able to reproduce standard adult-provided representations. diSessa (2004) promotes *meta-representational* competence as including the ability to 'invent or design new representations' (p. 293), to understand their purposes, and to explain and critique the adequacy of representations.

4.2.1 Student-Generated Representations

Children, and indeed adults, encounter many mathematical 'ideas' through natural interactions with the world around them. Young children encounter the idea of division as sharing when they evenly distribute sweets among their friends. Through these experiences, they form mental images and intuitive mathematical knowledge. These non-symbolic internal representations can be utilized to perform basic mathematical tasks even before conventional symbolic representations have been externally introduced. For example, a substantial body of research has shown that young children can utilize their 'sense of magnitude' to compare, add, and subtract quantities without the use of number of words or symbols (e.g., Batchelor, Keeble, & Gilmore, 2015; Gellistel & Gelman, 1992). Depending on a child's mathematical experiences and understandings, existing internal representations may either facilitate or hinder their perception of important information conveyed by standard external adult-provided representations. Since we can only infer children's internal representations from the external representations they produce (Goldin & Shteingold, 2001), investigating student-created representations may yield helpful insights into their mathematical thinking. Indeed, children's self-generated representations may more closely reflect their level of conceptual understanding than their responses to imposed conventional representations. Both English (2012) and diSessa (2004) noted that research surrounding the role of student-created representations in young children's mathematical conceptual development is limited and complex.

A naturally developing medium of representation for children is drawing. 'Drawing can be a window into the mind of a child' (Wolek, 2001, p. 215), particularly from around the age of four years when children begin to deliberately use drawing in iconic, symbolic, and even mathematical ways (Machon, 2013). Drawings are often seen as products or static artifacts, but representation is a dynamic process reflecting cognitive functions and emerging understandings (Goldin & Shteingold, 2001; Thom & McGarvey, 2015). MacDonald (2013) cautions that an adult can easily misinterpret a child's mathematical meaning-making and advocates seeking the child's narrative of their drawing. Diezmann and English (2001) found that children's drawings of diagrams during mathematical problem solving could assist in finding an appropriate solution. Recognizing the pedagogical potential inherent in children's drawing activity, with a focus on supporting the 'mathematization' of their representations, is a worthy skill for teachers to acquire.

Another form of representation that can be particularly useful to both learners and teachers is that of gesture—small spontaneous movements of the hands or arms (Logan, Lowrie, & Diezmann, 2014). When children talk and explain, they often 'express new knowledge in gestures before they express it in speech' (Alibali & Nathan, 2012), and so gesture can support communication by young learners with emerging mathematical vocabulary. Representational gestures seem to be particularly useful for spatial thinking and are often activated through the use of other representations such as drawing (Elia, Gagartis, & van den Heuvel-Panhuizen, 2014; Thom & McGarvey, 2015) and verbal description (Alibali & Nathan, 2012; Logan et al., 2014). There is also evidence that teachers' gestures can influence the thinking and gestures of young children (Alibali & Nathan, 2012), suggesting the potential of gesture as a deliberate pedagogical strategy (Elia et al., 2014).

4.3 Reflecting on Our Prior Research

To achieve a better understanding of the critical relationship between representations and the learning of mathematics, we believe it is necessary to explore two fundamental questions: (1) What role do student-generated representations play in their mathematical conceptual development? and (2) how does a child's mind create, interpret, and process representations? Although these questions are intertwined, they focus on two different perspectives of learning. To invoke an analogical mental image—consider each question to be the other side of the same coin. The first question, with its focus on conceptual development, implies a learning trajectory—an anticipated general pathway from 'not knowing' to 'understanding'—in which representation plays a critical role. The second question, with its focus on cognitive processing, implies that there are specific mental operations needed if we are to make effective use of representations for accessing the mathematics they signify. To illustrate the potential insights offered by pursuing each of these questions, we have selected two short episodes from previous studies for re-analysis, with each episode analyzed using one of these two perspectives.

The first episode is a brief incident drawn from a series of classroom studies that focused on children's early number development (Bobis, 2011; Bobis et al., 2005) and is framed within a selected model of conceptual development—the *Pirie–Kieren Dynamical Theory for the Growth of Mathematical Understanding* (Pirie & Kieren, 1989, 1994). Through this episode, we seek to illustrate the fundamental role that child-created representations play in moving a child from an 'experience' through 'mathematizing', toward a mathematical understanding (Question 1). We also show how a child's gestures can give a glimpse of an internal representation moments before it manifests as an external representation.

The second episode is a set of responses from two children, drawn from task-based interviews in a large study conducted to explore children's notions of probability (Way, 2003a, 2003b). The re-analysis of the representations used in the task, and the children's responses are based upon Duval's (1999, 2006) framework of 'cognitive architecture' for the representation and visualization of mathematics (Question 2). Through this episode, we seek to illustrate the complexity of representations commonly used by researchers (and teachers) and the difficulties faced by children when attempting to interpret such representations.

4.4 Episode 1: Emma: Covering the Desk with Books

Data reported in this episode were collected as part of research surrounding a system-wide numeracy professional development program (Bobis, 2011). This particular episode and accompanying work sample have not previously been targeted for deeper analysis, but were considered as part of a larger dataset relating to student numeracy outcomes (Bobis et al., 2005). In the following reporting and analysis, the episode is examined to highlight how, when, and why particular mathematical concepts and understandings emerged and developed during the actual drawing process.

4.4.1 The Task and Emma's Representation

Emma was a five-year-old child attending the final school term of kindergarten in a public school situated in the northern suburbs of Sydney, Australia. The mathematics lesson in which this episode occurred was conducted by her classroom teacher (Mrs. Baker) and observed by one of the chapter's researchers/authors. The researcher acted as a participant observer, interacting with students as they worked on tasks set by their teacher and taking field notes that included the recording of conversations between the teacher or researcher and the students.

Mrs. Baker began the lesson by introducing the idea that smaller shapes can be used to cover large areas. To help explain this idea, she placed five identical textbooks end-to-end along one length of a large table and eleven more books along the adjacent side of the same table, thus forming a partial array with five books along one side and twelve books along the other. As she placed each book on the table, Mrs. Baker emphasized that the books should not overlap each other or have any gaps between them. She commented, 'If we keep going, we can find out how many books we need to cover the whole area of this table.' Mrs. Baker removed the books from the table. She then provided a range of everyday equipment, including paper squares, ceramic tiles, and pattern blocks, and invited her students to cover various surfaces in their classroom without creating any overlaps or leaving gaps between the items.

The majority of children immediately started covering the surface of books, chairs, and tabletops using the supplied equipment. However, Emma was observed moving to the other side of the classroom where she retrieved some blank paper and a pencil. She sat at a nearby table and began drawing. The teacher moved closer to Emma and questioned her about her drawing:

Emma I want to know how many books will cover the table and there aren't enough (books) so I'll have to draw them.

Teacher Okay. I'll be interested to know what your answer is when you're finished.

Mrs. Baker moved back to assist other children who were busily working with the equipment. Emma continued drawing the arrangement of books from memory but looked back to the table where the books had originally been placed on two occasions, as if to remind her of what the arrangement looked like. First, she drew the column of rectangles (books) on the left side of the figure (see Fig. 4.1). She drew each rectangle one after the other, starting at the top and working her way down the column. She verbalized her counting of the rectangles as she drew them,

Fig. 4.1 Emma's diagram representing books covering a table

ensuring she had 5 books in total. Emma then wrote the numerals 1–5 in each rectangle, starting with the number '1' in the bottom rectangle.

Emma then drew the second column of books starting with the top rectangle. After drawing the first rectangle, she paused, slightly tilted her head to the side, and drew imaginary horizontal and then vertical lines 'in the air' about five centimeters above the paper with her pencil. These gestures were quickly followed by a quiet utterance, 'Ahh!'. Emma resumed drawing by extending the top and bottom horizontal lines of the entire rectangle. She then drew a series of intersecting vertical lines until there were 12 books across the bottom of the large rectangle, adjusting the length of her top and bottom horizontal lines on two occasions (after eight and 11 books) to ensure she could fit 12 books across the bottom row. Emma made numerous utterances while drawing, such as, 'mm..' and 'yes'.

She completed her diagram and looked at the page in silence for a few seconds. The following conversation then occurred between the researcher and Emma.

Researcher	Emma, tell me about your drawing.
Emma	It's the books. I'm trying to count how many we need to cover the table. I just put numbers in the books that Mrs. Baker used. All the others are just pretend because we don't have that many books… I just have to count all the squares now.
Researcher	How are you going to count them?
Emma	Mmm… I'm going to skip-count by fives.

Emma then pointed at the rectangle labeled '1' and proceeded to count 5, 10, 15, 20…, pointing to each numbered box across the bottom of the rectangle until she reached the twelfth box, simultaneously verbalizing '60'. She then wrote '60' next to her drawing. The researcher questioned Emma further:

Researcher	Why did you decide to count by fives?
Emma	You could also do it by twelves, but I don't know how to do that... fives are easier.
Researcher	Can you show me where each group of five books is on your drawing so I can see how you counted them?
Emma	(running her finger down the first column of rectangles) There's five... (roughly running her finger down each of the next three columns) five, five, five... all the way to the end...twelve times...there are twelve books in each row.

Emma then wrote '5 rows of 12' next to her diagram.

4.4.2 Re-Analyses of Emma's Representation

The re-analysis of this episode draws upon the Pirie–Kieren Dynamical Theory for the Growth of Mathematical Understanding (Pirie & Kieren, 1989, 1994) to explore and demonstrate the mathematical 'knowings' embedded in Emma's representation. The Pirie–Kieren theory consists of eight layers depicted as a set of nested rings that are labeled, from the center outward: primitive knowing, image making, image having, property noticing, formalizing, observing, structuring, and inventizing. The layers of knowing build from a learner's prior knowledge toward more sophisticated understandings and abstraction, but do not indicate a linear progression, because the learner will constantly cycle back (or 'fold back') to preceding layers to draw upon prior understandings to help them make sense of new knowledge and experiences. It is beyond the scope of this chapter to provide a comprehensive explanation of the entire Pirie–Kieren theory. However, a brief explanation of the first four levels of understanding—those that are most relevant to the current analysis—is provided prior to the theory being used to analyze aspects of the episode. While we cannot exactly know what a young child's internal 'knowings' are, we can gain insights from a child's movements, verbalizations, and their written and drawn representations (Thom & Pirie, 2006). It is through a consideration of each of these modes of representation that we come to more fully appreciate Emma's developing conceptual understandings.

4.4.2.1 Primitive Knowing

The first level in the process of mathematical understanding according to the Pirie–Kieren theory is called 'primitive knowing'. 'Primitive' refers to the starting place for growth of all other knowings. It does not simply mean immature understandings, but is everything a person already knows about the current task—both formal and informal knowledge. Emma brought to the task her knowledge of counting by ones to 12, her knowledge of numerals that assisted her to keep some order in her individual rectangles. The numeral recordings also allowed her to keep track of the number of books in a row and a column.

Emma seemed to be contemplating a method for counting all the rectangles when the researcher questioned her about a strategy for doing this. Prompted by this questioning, Emma appeared to have folded back to her existing knowledge (primitive knowing) of skip-counting by fives to help her find the total number of rectangles. She demonstrated a clear understanding of multiples of five when she quickly traced her finger down the first three columns of rectangles as part of her explanation about her strategy. Despite her recognition that it was also possible to count by 12s, it is not possible to determine from this interaction if she understood the commutative relationship between 'five rows of 12' and '12 columns of five.'

Emma also already knew how to draw individual rectangles side by side by using the side of one shape to form the side of a new rectangle. In this way, she was able to avoid gaps and overlapping rectangles.

4.4.2.2 Image Making

The Pirie–Kieren theory states that new knowledge is created when one performs some mental or physical actions. The idea being, such actions give rise to new images in a process of 'image making.' These images can occur in mental, oral, symbolic, pictorial, or physical form. The first evidence that Emma's understanding was evolving occurred when she gestured with her pencil above her drawing, as if testing the usefulness of vertical and horizontal lines to represent the remaining books. This moment seems to be quite pivotal in her ability to solve the problem as evidenced by her hand gestures, head movements, and utterances *(Ahh! Mmm…, etc.)*. These physical actions reflect Emma's mental processing as she started to make *new* sense of her diagram—even the parts that were yet to be drawn.

4.4.2.3 Image Having

At the 'image having' level, the learner achieves an initial degree of abstraction. She no longer needs physical actions to help make sense of an image. Instead, she is able to 'fold back,' drawing upon knowledge that is already known and apply it to the current activity. Emma folded back to her ability to skip-count by fives. She did not need to resort to the specific activities that gave rise to her understanding of skip-counting in the first place, but could effortlessly apply this knowledge to derive her answer (or image) of 60.

4.4.2.4 Property Noticing

The Pirie–Kieren theory proposes that once images are effectively understood, learners are then ready to make connections and distinctions among new and previously held images. It is through this 'property noticing' that learners are able to reflect on existing understandings so as to extend it even further. Stimulated by the researcher's questions, Emma reflected on the fact that there were five rows of 12 (and 12 lots of five). She then made a generalization and wrote '5 rows of 12.' However, there is still no clear evidence from this episode that Emma understood the connection that '5 rows of 12' books is the same as 60 books. From the data collected during this episode, it could not be determined whether Emma's understanding moved beyond property noticing.

Examining multiple aspects of her diagram's evolution in such detail, including Emma's accompanying gestures and utterances, not only reveals insights into how she solved the problem, but the shifts in her thinking during the actual drawing process. As suggested by Thom and McGarvey (2015), the representation and mathematical thinking coevolved. Emma's diagram was not just a re-representation of what she knew, but the act of drawing was an important stimulus to furthering her mathematical conceptual understanding.

4.5 Episode 2: Albert and Jessica: Comparing Sample Spaces

As part of a study assessing children's notions of randomness and chance, game-like tasks involving two different random generators were used during individual interviews (Way, 2003a, 2003b). Children were asked to make choices and explain their thinking about drawing small colored objects (bears) from a box and playing a car-race game with various spinners. The youngest children (5–6 years old) responded very differently to the two materials, more readily perceiving differences in likelihood when using the spinners. A further task is attempted to determine whether the children could perceive the mathematical equivalence between the two sample spaces, one represented by quantities of discrete items (bears) and the other by areas (sectors on a spinner).

4.5.1 The Task and Responses of Albert and Jessica

The task scenario was: We want to play the car-race game but have lost the spinners. Could we put some bears into the box, so that the color drawn out would tell us which car to move forward?

In the interview, the child was shown, in turn, Spinner A with equal parts (red, yellow, blue, green) and Spinner B with unequal parts (see Fig. 4.2), then asked to create their own matching 'sample space' with the bears.

Almost all the children were able to design two versions of bears-in-the-box for the 'equally likely' Spinner A, such as one of each color, or three of each color.

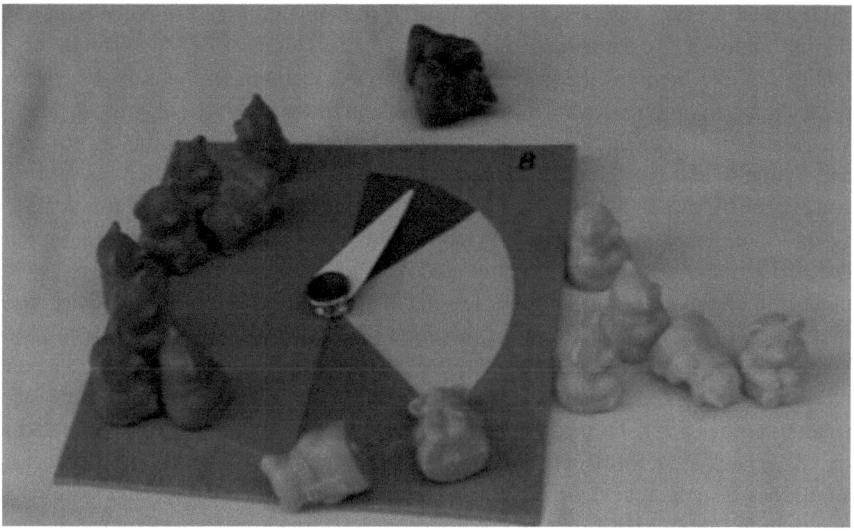

Fig. 4.2 Spinner B (Ratio—4 red: 2 yellow: 1 blue: 1 green) with equivalent quantities of bears

However, Spinner B was a challenge. Albert's response (see following excerpt) typifies half of the twelve interviewees, and Jessica's response illustrates the extent of the remaining six responses.

Albert (5 years 7 months), in response to the Spinner B task, put one of each color bear in the box and did not want to change this even after some prompting by the researcher.

Jessica (5 years 11 months), in response to the Spinner B task, lined up four bears along the edge of one of the sectors on the spinner—a radius of the circle. She placed four of each color bear into the box and explained, '*I counted it up from down to up.*' When asked whether the box would be likely to produce the same winner as the spinner, she made an attempt to relate the size (area) of some sectors to the number of bears of that color. Jessica said, '*I think I'll put some more extra reds in—so Red should win, but I don't care about the others—but Yellow looks like it should have a bit more.*' The final contents of the box were 6 red, 4 yellow, 4 blue, 4 green.

4.5.2 Re-Analysis of the Task and Responses

At the time of the study, the aim of this task was to elicit children's conceptual development regarding basic concepts of probability. To shift the focus to the cognitive processing of the representations used in the task, we re-analyzed the episode using Duval's theoretical model for the cognitive architecture of mathematical activity (Duval, 1999, 2006). Duval advocates just such an approach, saying that, '... any task or problem that the students are asked to solve requires a double analysis, mathematical and cognitive...' (Duval, 1999, p. 24). In brief, Duval's model involves the premise that to access mathematics at a deep level, we must create, manipulate, and interpret representations, such as diagrams, graphs, verbal statements, or number relations. These representations are structures of signs and symbols and are therefore *semiotic representations*. For representations to be comprehensible and useful, we must have sufficient knowledge of the semiotic system that determines the signs, symbols, and conventions for their use (e.g., the decimal number system or projective geometry). There are many semiotic systems used in mathematics, and so there are many possible representations of the same mathematical idea. The representations might be held mentally (internal) or made visible (external), with constant interplay between the two forms. Some representations are more complicated because they utilize more than one semiotic system (like a number line). To meaningfully work with representations, we often need to translate back and forth between two or three semiotic systems—like modeling addition with concrete materials, verbalizing the number relation, and writing an equation (Rivera, 2014), or moving between 2-D drawings and 3-D constructions (Thom & McGarvey, 2015). According to Duval, '...the ability to change from one representation system to another is very often the critical threshold for progress in learning and for problem solving' (2006, p.107).

4.5.2.1 Re-Analysis of the Task Representations

Playing the bears-in-a-box and the car-race games gave the children opportunity to experience the mathematical phenomenon of chance events. The experience was mediated through two different physical representations of sample space, constructed by the researcher.

The original bears-in-a-box game used three yellow bears and one red bear, which were shown, then hidden inside the box, thus prompting the children to work with a perceptual mental image of the bears. Drawing on Duval's proposition that a person's interpretation of a representation is dependent on their existing semiotics systems allows us to hypothesize a list of possible interpretations. The children could have interpreted the image of the bears in a range of ways, including:

- general impression of 'more yellow than red';
- iconic representation of the numbers 3 and 1;
- translation of the image to a semiotic representation depicting a specific part-whole situation with three-quarters yellow and one-quarter red;
- translation of the image to a semiotic representation depicting the ratio 3:1, perceived as representing any sample space with this proportion.

The original car-race game utilized two separate representations created by the researcher—the racetrack (with four colored disks representing cars) and the spinners as shown in Fig. 4.3. As the racetrack was not used in the follow-up task central to the episode, it will not be re-analyzed here.

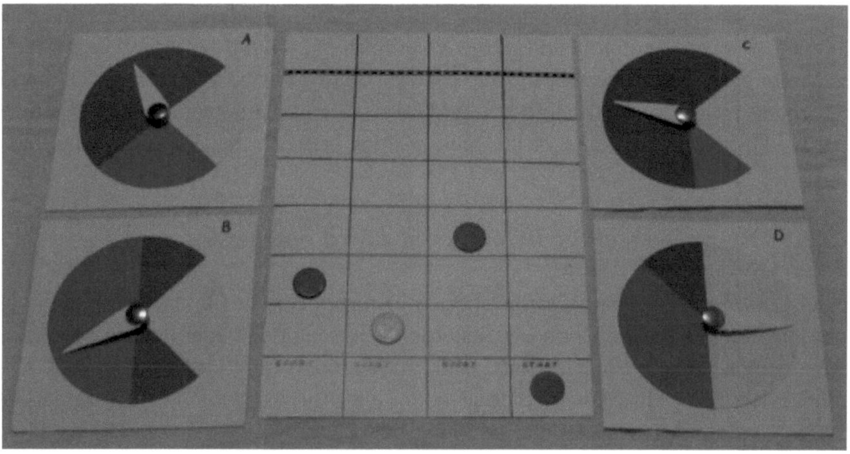

Fig. 4.3 Material for the 'car'-race game

During each game, the relevant spinner remained visible to the child, removing the cognitive demand of recalling the perceptual image. The image of the spinner can be interpreted in a range of ways, including:

- general impression of the 'size' of each color allowing a simple direct comparison of color portions;
- translation of the image to a semiotic representation depicting a specific part-whole situation, with individual fractions determined either through estimation of area, recognition of familiar depictions (e.g., that's what ¼ looks like), reasoning (e.g., half of a half is one-quarter), or a combination of strategies;
- extension of the fraction representation to determine ratios, using the smallest part as the unit fraction of one-eighth as a starting point;
- translation of the image to the geometric representation of a circle, with sectors determined by interior angles, etc.; and
- the association might also be made with sector graphs and percentages, which draw on other semiotic systems.

The re-analysis of the representations of sample spaces designed by the researcher reveals the cognitive complexity of interpreting these commonly used mathematical tools. To move beyond an initial visual image, the children needed to 'mathematize' the image by transforming it into a semiotic representation. To accomplish this translation, the children needed to have familiarity with an appropriate semiotic system and recognize the key mathematical features of the image to transform. The focus task for this episode—putting bears in the box to match a spinner—challenged the children to compare the two different representations. If the children had interpreted both representations using the same semiotic system (e.g., fractions), then the relationship between the two materials would be relatively straightforward. If each of the representations was interpreted via different semiotic systems (e.g., whole numbers and fractions, or whole number and circle geometry), then comprehending the relationships would be more cognitively demanding. In a third scenario, the children may only be operating with initial perceptual images and therefore struggle to make any mathematical interpretations that would help make sense of the relationships between the two representations.

4.5.2.2 Re-Analysis of the Responses of Albert and Jessica

Using the bears to create a sample space to match Spinner B is depended on an interpretation of the two imposed external representations. The children's verbal explanations and physical actions with the materials provided some observable indications from which their internal cognitive processing of the representations could be inferred.

Albert attended only to the color property of the spinner, recognizing the need to include each of the four colors when putting bears in the box. It appears he could not move beyond the simple visual image he held for each of the materials and could not perceive them as mathematical representations. Albert's responses to earlier tasks suggest he may have had some sense that differing quantities of color mattered, but he was not able to utilize that observation as a basis for a comparison strategy in this task.

Jessica interpreted the bears as an iconic representation of number. That is, she could count the bears for each color and work directly with those numbers. This indicates a first step toward mathematizing the image she had of the bears-in-box material and translation it into another semiotic representation. (we say 'first step' because there is limited evidence that she could use this representational system to access the underlying probability concepts) She attempted to transfer this thinking to Spinner B. She chose one structural feature of the spinner—a dividing line between color sectors (actually the radius of the circle)—and lined up some bears along its length. As four bears fitted along the line, she proposed that four of each color needed to go into the box. Jessica seemed to be attending to the size of color portions on the spinners in previous tasks, and when prompted to review her solution, she returned to this perception. Seeing that the red portion of the spinner was larger than the other colors, she added two more red bears. Jessica's words and actions suggest that she attempted to reconcile the two representations by looking for some feature on the spinner to 'measure' with 'unit bears' and produce numbers by counting the bears.

Examining this episode from the perspective of the cognitive processing of representations reveals the mathematical complexity of two materials commonly used in games and classroom activities. This perspective highlights the need to deeply consider the cognitive demands of task designs, in terms of the semiotic systems that define particular representations. We see Albert and Jessica at the beginning of their long development in learning the many semiotic systems they will need to comprehend mathematics and wonder how often young children are baffled by the mathematical representations that adults present to them.

4.6 Current Research Directions

Our current research is driven by the proposition that representation is a dynamic process integral to mathematical thinking, and that by deepening our understanding of the complexities of mathematical representation, we can inform effective pedagogy—the outcome being enhanced learning for children. One of the perplexing difficulties in such research is the enigmatic nature of internal images and thoughts. This leads to the premise that understanding children's self-generated external representations, in all their observable forms, provides indications of their internal representations. As expressed by Goldin and Shteingold (2001),

> The interaction between internal and external representation is fundamental to effective teaching and learning. Whatever meanings and interpretations the teacher may bring to an external representation, it is the nature of the student's developing internal representation that must remain of primary interest (p. 2).

In our theoretical work, as indicated by the re-analysis of the episodes with children presented in the chapter, we have begun 'testing' theories of cognition, conceptual development, and growth of understanding for their capacity to deeply explain the role of representation in children's learning of mathematics. As this work progresses, we are finding that probing the role of representations more deeply in theoretical models, such as the *Pirie–Kieren theory* (Pirie & Kieren, 1989, 1994) and Duval's model (1999), reveals the potential to extend and elaborate the theory, particularly regarding young children's learning.

The *dynamic* nature of young children's representation and the ways in which their thinking appears to be created, combined, and expressed through multiple representational modes, such as imagining, drawing, verbalizing, moving, emoting, and gesturing, has drawn us toward the theory of *embodied cognition*. The theory of embodied cognition differs from other cognitive theories in its recognition that, because we interact with the environment through our entire bodies, 'thinking' and learning involve all of our 'being' not just the mind (Stillman, 2014; Wilson, 2002). We believe that this view affords great potential for noticing and understanding the ways in which young children create internal representations and express them, perhaps simultaneously, as external representations (Thom & McGarvey, 2015), as was seen in Emma's gesture and diagram in the earlier episode.

Our empirical research has just commenced and has a triple focus:

- the child's internal–external representation processes—here our initial studies are exploring the role of gesture, as well as developing methodologies for a more holistic interpretation of representational processes;
- the child's interaction with teacher representation—here our main concern is with the 'conceptual distance' between child-created representation and teacher imposed (conventional) mathematical representations; and
- the teacher's own understanding and use of representations as part of their 'knowing' of mathematics—here we extend our previous research, which

reveals the transformative potential of targeting a teacher's own representational repertoire (Way, Bobis, Anderson, & Cameron, 2013).

These foci lead us to consider the important impact of teacher intervention in supporting students during the physical generation and sense-making stages of creating their representations. First, we encourage teachers to actively seek children's own mathematical representations, rather than solely providing them with conventional and adult-produced ones. Second, education authorities need to ensure that teachers are capable of noticing and interpreting the mathematical structure and properties inherent in children's representations. Third, we highlight the importance of teacher intervention in helping make the mathematics inherent in student representations explicit, and of helping the children make connections between multiple representations of the same concept.

4.7 Conclusion

In this chapter, we have highlighted that for teachers to better support children in their learning of mathematics, they need to attend closely to the development of their representations. Moreover, we not only emphasize the importance of taking notice of the finished 'products' but the critical importance of attending to the entire process involved in generating the representations—including the accompanying movements and utterances. It is only through attending to the dynamic process in its entirety that we can develop a deeper understanding of children's representations and, in doing so, be better prepared to support them make explicit connections between their representations and mathematical concepts.

References

Alibali, M., & Nathan, M. (2012). Embodiment in mathematics teaching and learning: Evidence from learners' and teachers' gestures. *Journal of Learning Sciences, 21*(2), 247–286.

Batchelor, S., Keeble, S., & Gilmore, C. (2015). Magnitude representations and counting skills in preschool children. *Mathematical Thinking and Learning, 17*(2–3), 116–135.

Bobis, J. (2011). Mechanisms affecting the sustainability and scale-up of a system-wide reform. *Mathematics Teacher Education and Development, 13*(1), 34–53.

Bobis, J., Clarke, B., Clarke, D., Thomas, G., Wright, R., Young-Loveridge, J., et al. (2005). Supporting teachers in the development of young children's mathematical thinking: Three large scale cases. *Mathematics Education Research Journal, 16*(3), 27–57.

Braswell, G. (2015). Observations of representational practices by Indian-descent children in a US preschool classroom: Connections among people, spaces and artifacts. *Early Childhood Education Journal, 43*(2), 135–142.

Chao, S., Stigler, J., & Woodward, J. A. (2000). The effects of physical materials on kindergartners' learning of number concepts. *Cognition and Instruction, 18*(3), 285–316.

Diezmann, C., & English, L. (2001). Promoting the use of diagrams as tools for thinking. In A. Cuocu & F. Curcio (Eds.), *The role of representation in school mathematics: 2001 Yearbook of the National Council of Teachers of Mathematics* (pp. 77–89). Reston, VA: NCTM.
Duval, R. (1999). Representation, vision and visualization: Cognitive functions in mathematical thinking: Basic issues for learning. In F. Hitt & M. Santos (Eds.), *Proceedings of the 21st North American PME conference* (pp. 3–26). Cuernavaca, Morelos, Mexico.
Duval, R. (2006). A cognitive analysis of problems of comprehension in a learning of mathematics. *Educational Studies in Mathematics, 61*(1–2), 103–131.
Elia, I., Gagartis, A., & van den Heuvel-Panhuizen, M. (2014). The role of gestures in making connections between space and shape aspects and their verbal representations in the early years: findings from a case study. *Mathematics Education Research Journal, 26,* 735–761.
English, L. (2012). Young children's metarepresentational competence in data modeling. In J. Dindyal, L. P. Cheng & S. F. Ng (Eds.), *Mathematics education: Expanding horizons* (Proceedings of the 35th annual conference of the Mathematics Education Research Group of Australasia, pp. 266–273). Singapore: MERGA.
diSessa, A. (2004). Metarepresentation: Native competence and targets for instruction. *Cognition and Instruction, 22*(3), 293–331.
Gallistel, C. R., & Gelman, R. (1992). Preverbal and verbal counting and computation. *Cognition, 44,* 43–74. doi:10.1016/0010-0277(92)90050-R.
Ginsburg, H. P., Lee, J. S., & Boyd, J. S. (2008). Mathematics education for young children: What it is and how to promote it. *Social Policy Report of the Society for Research in Child Development, 22,* 3–23.
Goldin, G., & Shteingold, N. (2001). Systems of representation and the development of mathematical concepts. In A. Cuoco (Ed.), *The roles of representations in school mathematics, NCTM 2001 Yearbook* (pp. 1–23). Reston VA: NCTM.
Gravemeijer, K. (2009). How emergent models may foster the constitution of formal mathematics. *Mathematical Thinking and Learning, 1*(2), 155–177.
Logan, T., Lowrie, T., & Diezmann, C. (2014). Co-thought gestures: Supporting students to successfully navigate map tasks. *Educational Studies in Mathematics, 87*(1), 87–102.
MacDonald, A. (2013). Using children's representations to investigate meaning-making in mathematics. *Australasian Journal of Early Childhood, 38*(2), 65–73.
Machon, A. (2013). *Children's drawings: The genesis and nature of graphic representation: A developmental study*. Madrid: Fibulas.
McNeil, N., & Jarvin, L. (2009). When theories don't add up: Untangling the manipulatives debate. *Theory into Practice, 46*(4), 309–316.
Meyer, M. (2001). Representation in realistic mathematics education. In A. Cuoco (Ed.), *The roles of representations in school mathematics, NCTM 2001 Yearbook* (pp. 238–250). NCTM: Reston VA.
Moyles, J., Adams, S., & Musgrove, A. (2006). Early years practitioners' understanding of pedagogical effectiveness: Defining and managing effective pedagogy. In R. Parker-Rees & J. Willan (Eds.), *Early years education—major themes in education: Vol. 3. Policy and practice in early education and care* (pp. 306–322). London: Routledge.
Obersteiner, A., Reiss, K., Ufer, S., Luwel, K., & Verschaffel, L. (2014). Do first graders make efficient use of external number representations? The case of the twenty-frame. *Cognition and Instruction, 32*(4), 353–373.
Papic, M. (2013). Improving numeracy outcomes for young Australian Indigenous children through the Patterns and Early Algebra Preschool (PEAP) Professional Development (PD) Program. In L. English & J. Mulligan (Eds.), *Reconceptualizing early mathematics learning* (pp. 253–281). NY: Springer.
Papic, M., Mulligan, J. T., & Mitchelmore, M. C. (2011). Assessing the development of preschoolers' mathematical patterning. *Journal for Research in Mathematics Education, 42*(3), 237–268.
Pirie, S., & Kieren, T. (1989). A recursive theory of mathematical understanding. *For the Learning of Mathematics, 9*(3), 7–11.

Pirie, S., & Kieren, T. (1994). Growth in mathematical understanding: How can we characterise it and how can we represent it? *Educational Studies in Mathematics, 26,* 61–86.
Rivera, F.D. (2014). From math drawings to algorithms: emergence of whole number operations in children. *ZDM—The International Journal on Mathematics Education, 46*, 59–77.
Steenpass, A., & Steinbring, H. (2014). Young students' subjective interpretations of mathematical diagrams—elements of the theoretical construct "frame-based interpreting competence". *ZDM—The International Journal on Mathematics Education, 46*(1), 3–14.
Stillman, G. (2014). Views of cognition: Different lenses for 'looking in' on classrooms. *ZDM—The International Journal on Mathematics Education, 46*, 493–496.
Thom, J., & McGarvey, L. (2015). The act of drawing(s): Observing geometric thinking with, in, and through children's drawings. (2015). *ZDM—The International Journal on Mathematics Education, 47*, 464–481.
Thom, J., & Pirie, S. (2006). Looking at the complexity of two young children's understanding of number. *Journal of Mathematical Behavior, 25,* 185–195.
Way, J. (2003a). The development of children's reasoning strategies in probability tasks. In Bragg, L., Campbell, C., Herbert, G. & Mousley, J. (Eds.), *Mathematics Education Research: Innovation, Networking, Opportunity* (Proceedings of the Twenty-sixth Annual Conference of the Mathematics Education Research Group of Australasia, pp. 736–743). Geelong, Vic.: MERGA.
Way, J. (2003b). *The development of children's notions of probability.* (Unpublished doctoral dissertation) University of Western Sydney. uwsprod.uws.dgicloud.com.
Way, J., Bobis, J., Anderson, J., & Cameron, L. (2013). Representations of concepts as a catalyst for change in teacher pedagogical content knowledge. *Procedia: Social and Behavioral Sciences, 106*, 2248–2258.
Wilson, M. (2002). Six views of embodied cognition. *Psychonomic Bulletin & Review, 9*(4), 625–636.
Wolek, K. (2001). Listen to their pictures: An investigation of children's mathematical drawings. In A. Cuoco (Ed.), *The roles of representations in school mathematics, NCTM 2001 Yearbook* (pp. 215–227). Reston VA: NCTM.

Authors Biography

Professor Janette Bobis is a mathematics educator and researcher at the University of Sydney, Faculty of Education and Social Work. Drawing upon her experience as a primary school teacher, she lectures in primary mathematics education and conducts research into teacher's mathematical knowledge, beliefs, and practices. She also studies the impact of these teacher characteristics on early childhood, primary and middle year students' understandings of mathematics, and their motivation and engagement in mathematics.

Associate Professor Jennifer Way is a mathematics educator and researcher at the University of Sydney, Faculty of Education and Social Work. A central theme of her work across early childhood and primary years has been conceptual development in children and pedagogical techniques that support children's mathematics engagement and understanding. A particular focus has been investigating the roles that various forms of representations, including digital interactives, play in learning and teaching mathematics.

Chapter 5
Geometry Learning in the Early Years: Developing Understanding of Shapes and Space with a Focus on Visualization

Iliada Elia, Marja van den Heuvel-Panhuizen and Athanasios Gagatsis

Abstract In this chapter, we address children's geometry learning in the early years with a focus on visualization. We start the chapter with some background information about visualization in mathematics and geometry and its relationship with language and gestures paying special attention to the early years. The next parts of the chapter aim to give insight into how young children solve geometrical activities with emphasis on the uses of visualization in developing understanding of space and shape concepts. In particular, we discuss three different research approaches which investigated young children's development of geometrical thinking when dealing with shapes' transformations, imaginary perspective taking, and space and shape aspects with the use of gestural and verbal acts. Finally, in light of the above, a number of conclusions are drawn about the multiple qualities and uses of visualization in the development of the understanding of shapes and space and the diverse factors that may intervene in early geometry learning which involves the use of visualization.

Keywords Geometry · Visualization · Shape transformation · Imaginary perspective taking · Spatial concepts · Gesture · Early years

I. Elia (✉) · M. van den Heuvel-Panhuizen
Freudenthal Institute of Utrecht University, Utrecht, Netherlands
e-mail: I.Elia@uu.nl

M. van den Heuvel-Panhuizen
e-mail: m.vandenheuvel-panhuizen@uu.nl

A. Gagatsis
University of Cyprus, Nicosia, Cyprus
e-mail: gagatsis@ucy.ac.cy

V. Kinnear et al. (eds.), *Forging Connections in Early Mathematics Teaching and Learning*, Early Mathematics Learning and Development,
https://doi.org/10.1007/978-981-10-7153-9_5

5.1 Introduction

Geometry is an indispensable part of contemporary early childhood curricula and educational programs (e.g., Sarama & Clements, 2009). This mathematical domain includes "the study of spatial relationships of all kinds; relationships that can be found in the three-dimensional space we live in and on any two-dimensional surface in this three-dimensional space. These relationships can be discovered all around us" (Egsgard, 1970, p. 478). This approach to geometry is in line with the view which is adopted in this chapter about early geometry, that it needs to be studied as a subject with a dynamic, spatial, and imaginative character, rather than as a static subject focusing on the naming and sorting of shapes (Moss, Hawes, Naqvi, & Caswell, 2015).

Researching geometry learning and teaching in the early years has received increasing attention in an attempt to promote the importance of geometry in early childhood and to propose new directions (Sinclair & Bruce, 2015). Nevertheless, there is still much to be done to unravel young children's development in this domain (Dindyal, 2015). This chapter addresses young children's geometrical thinking with a focus on visualization, which Duval (2014) considers "the first crucial point in geometry learning" (p. 1). Visualization is an inherent component of students' making sense of aspects of space and shape when learning and thinking in geometry. For example, visualizing geometrical shapes and their manipulation can be a powerful heuristic for solving geometrical problems.

5.2 Visualization

Visualization has been defined and used in various ways in the literature. In this chapter, we adopt the broad meaning provided by Arcavi (2003) for this term.

> Visualization is the ability, the process, and the product of creation, interpretation, use of and reflection upon pictures, images, diagrams, in our minds, on paper or with technological tools, with the purpose of depicting and communicating information, thinking about and developing previously unknown ideas, and advancing understandings. (p. 217)

Mental images, external representations, visualization processes, and abilities are major constituents of visualization (Gutiérrez, 1996). A mental image is the basic component in visualization and refers to "any kind of cognitive representation of a mathematical concept or property by means of visual or spatial elements" (Gutiérrez, 1996, p. 9). According to Presmeg (1986), mental images (of high school learners) can be classified into five types: (a) concrete images ("picture in the mind"), (b) kinaesthetic images (of physical movements), (c) dynamic images (images in the mind that are moved or transformed), (d) pattern images (visual representations of abstract relationships), and (e) images of formulae (mental images with symbols as they appear, e.g., in textbooks). Presmeg (2006) points out that the categories of mental images may overlap and that the concrete images need to be articulated with analytical thinking to be effectively employed in mathematics.

Mental images are generated and transformed on the basis of the interpretation of external representations or objects and are expressed in verbal or graphical form, so there is an interaction between mental images and external representations (Yakimanskaya, 1991). External representations which may support visual reasoning in mathematics include, for example, pictures, diagrams, drawings, and verbal descriptions standing for mathematical concepts or properties (Gutiérrez, 1996).

In visualization, two key processes of dealing with visual images are performed either mentally or physically: interpreting/transforming information into visual terms (visual processing) and interpreting visual images to produce information (interpretation of figural information) (Bishop, 1983; Gutiérrez, 1996). In performing the above processes, there are various visualization abilities that need to be developed and used, depending on the task and the images concerned. According to Gutiérrez (1996), major abilities are, for example, figure–ground perception (identifying figures out of a complex configuration), perceptual constancy (ability to focus on critical attributes of objects despite changes in other non-critical attributes, e.g., size, orientation), mental rotation, perception of spatial positions (e.g., relate an object to oneself), perception of spatial relationships (e.g., relate several objects to one another or to oneself), visual discrimination (compare objects, images, etc., to one another).

5.3 Visualization in Geometry

Like any other mathematical object, objects in geometry are abstract concepts and can be studied only through their semiotic representations. A geometrical figure is a visual representation constructed with specific tools (e.g., ruler, compass, software). A figure integrates semiotic representations which are produced within three different registers: (a) shapes, (b) magnitudes, and (c) marks and words. In a geometrical task which includes figures, three corresponding types of cognitive activity are required: seeing shapes and recognizing what is seen, measuring magnitudes and comparing, and making inferences from the given properties represented in marks and words (Duval, 2014). Given that figures possess a central role in geometrical tasks and activities, visualization is a key cognitive process of geometrical thinking. Visualization involves the recognition of figural units which can be identified in a configuration of shapes as well as operating with figures (Duval, 1995). In geometry, visualization goes beyond a "spontaneous perceptual way of seeing" a figure (Duval, 2014, p. 11). This visual perception applies to any visual representation of material objects or spatial organization (images, diagrams, plans, etc.) outside mathematics. Perceptual shape recognition is sometimes misleading for the recognition of geometrical properties and, therefore, for the recognition of the geometrical objects represented (object recognition) (Duval, 2014). On the contrary, visualization enables the simultaneous and immediate apprehension of a configuration as a whole by distinguishing at the same time all its figural units, that is, all the elements that can be visually identified in a figure (characterized by their dimensionality, for example, 2D shapes or 1D objects, i.e., lines and segments), and their interrelations and also by recognizing other possible configurations. This is a mathematical way of looking at a figure which enables learners to develop the understanding that a geometrical figure is a representation of the abstract geometrical object and its properties, and not the perceptual object. This understanding, which plays an essential role in reasoning, defining, and problem solving or proving in geometry (Duval, 2014), is difficult for children to grasp, particularly for young children (Dindyal, 2015).

Visualization is the basis for the heuristic use of figures in geometry problems, as it enables the solution of problems without explicit references to properties. This is based on transformations of the 2D figural units into other figural units of the same dimensionality (2D), which Duval (2014) called the "operative apprehension of figures" (p. 15). More specifically, operative apprehension is a form of visual processing on geometrical figures which depends on the various ways of modifying a given figure, including the mereologic way, the optic way, and the place way. The mereologic way refers to the division of the whole given figure into parts and the combination of them in another figure or subfigures (reconfiguration). The optic way is when one makes the figure larger or narrower, or slant, while the place way of modification refers to its position or orientation variation. Each of these different modifications can be performed mentally or physically, through various operations. Within the operative apprehension, the given figure becomes a starting point to explore other configurations that stem from the applications of these visual operations (Duval, 1995).

The above theoretical considerations were focused on visualization of plane shapes. Considering that geometry refers to a model of space (Soury-Lavergne & Maschietto, 2015), spatial visualization, and reasoning is prevalent in this mathematical domain and it underlies most geometrical thinking (Dindyal, 2015; Van den Heuvel-Panhuizen & Buys, 2008). This spatial interpretation of geometry is the natural way in which young children encounter geometry. In line with this view, Freudenthal (1973) was one of the pioneers who argued strongly for starting with spatial geometry in the early years of education: "Geometry is grasping space… that space in which the child lives, breathes and moves. The space that the child must learn to know, explore, conquer, in order to live, breathe and move better in it." (p. 403) Spatial reasoning and visualization abilities develop through children's activities. For example, when playing hide-and-seek, children try to hide in a place in which they will be invisible to the child that is looking for them. Therefore, they try to imagine or to reason what the other child will and will not be able to see from different points of view.

Learning to mentally take a particular point of view and to see a model of it is an important aspect of spatial reasoning in the domain of geometry that is strongly highlighted by the NCTM Principles and Standards of School Mathematics (2000) in the K − 2 grades and also in the TAL teaching/learning trajectory for the first and second year of preschool (K1 and K2) (Van den Heuvel-Panhuizen & Buys, 2008).

5.4 Language, Gestures, and Visualization in Geometry

Geometrical figures alone are not sufficient for using properties to solve geometrical problems. Language production, either implicit or explicit, has an essential role and needs to be coordinated with visualization in doing geometry (Duval, 2014). According to Duval (2014) in geometry, three types of language production are necessary: (a) definitions of objects and properties, which are visualized through geometrical figures or configurations and theorems; (b) descriptions of the production of figures and configurations; and (c) inferences from given properties resulting in justifications or proofs. Among the three types of language production, definitions and descriptions are most closely associated with visualization (and this relationship is based on figural units rather than on whole figures).

Besides visual abilities, verbal skills need to be addressed and developed in the teaching and learning of geometry (Hoffer, 1981), beginning in early years of education (Dindyal, 2015). In geometrical activities (involving either 2D or 3D shapes), young children often share their experiences and reflections with others and thus develop their abilities to describe visual images (e.g., geometrical figures or configurations), spatial concepts, relations, and reasoning (Van den Heuvel-Panhuizen & Buys, 2008). By using language related to geometrical activities, children learn how to use geometry language which can support spatial visualization and reasoning. Developing children's geometry language means that they develop their knowledge and understanding not only of geometrical terms for shapes—"condensations of definitions" (Duval, 2014, p. 17)—but also of naming and describing actions and transformations that are performed with shapes, figures, and other objects, such as rotating, moving, and identifying their position—descriptions of configuration productions (Duval, 2014). It is to be noted that a broad perspective is taken here when using the term language, since apart from words, an indispensable component of children's communication about space and shapes is gesture (e.g., Elia, Gagatsis, & Van den Heuvel-Panhuizen, 2014).

The study of gestures in mathematical thinking, learning, and teaching has gained increased attention in mathematics education research in the last few years. The connection of gesture and visualization, acknowledged in the work of Presmeg (2006, 2014), examined the teachers' and students' visual thinking in high school mathematics classes. The use of gesture by a teacher or a student was considered a strong indication of visual imagery in mathematics learning and teaching. Gestures can serve as a dynamic representational tool of various mathematical ideas through which people can get a deeper level of consciousness of their meaning. The embodied character of gestures may facilitate the process of reaching abstract concepts through the visual modality. Moreover, as a consequence of this process, people can communicate mathematical concepts more easily (Nemirovsky & Ferrara, 2009). Thus, the roles of gestures in mathematical visualization are an issue of major importance. Further systematic investigation is still needed however (Presmeg, 2006; 2014), and this is particularly true for visualization in geometry, a domain in which the study of gestures has received limited attention, especially in the early years (Elia, Gagatsis, & Van den Heuvel-Panhuizen, 2014).

Gestures and spatial thinking are closely linked to one another, with gestures playing a significant role in cognitive processing (Alibali, 2005) and in conveying spatial information (McNeill, 1992) such as location and movement. The visuospatial nature of gesture makes it suitable for capturing spatial information. Gestures represent spatial properties and action-based characteristics of concepts (Krauss, Chen, & Gottesman, 2000). They assist speakers in activating mental images and in maintaining these spatial representations in working memory (Alibali, 2005). At the same time, using gestures to express spatial properties may support the activation of related mental representations of the concepts also in verbal form (Krauss, Chen, & Gottesman, 2000). As a result, visualization is used and manifested in diverse ways in geometry.

The next section of the chapter aims to give further insight into how young children solve geometrical activities, paying attention to the uses of visualization by children in developing understanding of space and shape concepts. In particular, we discuss three different research approaches drawn on in our studies in which visualization is used in different types of shape and spatial tasks, contexts, and with different tools and resources. Since mental images can be perceived only through external support (e.g., verbal, graphical) (Sutherland, 1995), in order to identify children's mental images and visualization abilities in these studies, we interpreted the actions produced or the outcomes of these actions that were the result of children's activity with mental images in response to the geometrical tasks.

5.5 A Dynamic Approach to Plane Geometry: Optic Transformation of Geometrical Figures

Identifying, describing, and classifying two-dimensional shapes have been the focus of a major body of research on geometry learning in mathematics education literature in the last years. Yet these aspects cover only a part of children's thinking in plane geometry. This focus probably stems from the strong and lasting influence of the Van Hiele (1985) model on mathematics education research at all age levels (Sinclair & Bruce, 2015). Taking a different perspective, Castelnuovo (1972) points out that children do not easily observe figures when they are steady, but rather when they move or vary in a continuous manner. Owens (1999) suggests that students visualize movement as they make connections between shapes. An example is constructing a square that becomes a rectangle as it gets thinner. Such intuitive thinking, which involves a continuous variation process, is called "dynamic intuition" (Castelnuovo, 1972). According to Presmeg (1986), it is a kind of dynamic visualization, as it involves the creation of dynamic mental images for specific geometrical figures.

Following this dynamic approach to geometry learning, Gagatsis, Sriraman, Elia, and Modestou (2006) explored young children's geometrical thinking by focusing on the dynamic use of drawings. Specifically, the study investigated the strategies children use in the optic transformation of 2D shapes (e.g., increasing or decreasing the size of figures) (Duval, 1995) and the relationship of these strategies with children's age and shape identification abilities.

The participants of the study were 291 children ranging from four to eight years of age. Children were asked to draw a stairway of triangles, squares, and rectangles, respectively, each one bigger than the preceding one, and then a stairway for each shape, each one smaller than the preceding one. To assess children's shape identification ability, children were asked to identify and color the squares, rectangles, and triangles among other figures. Three models of action/transformation strategies were observed in children's responses to the transformation tasks:

(a) Conservation of shape by increasing or decreasing both dimensions of a plane figure at the same time (Figs. 5.1 and 5.2). This was named as T-strategy (modification of Two-shape dimensions).
(b) Differentiating mainly one dimension of the figures: In the case of triangles, this dimension was the altitude to the base; that is why children sometimes produced isosceles triangles although their paths started with an equilateral one (most common case) (Fig. 5.3). In the case of rectangles, it was usually the longer side (Fig. 5.4); sometimes, a square occurred among the rectangles in a very natural way (Fig. 5.4). In the case of squares, children produced rectangles (Fig. 5.5). This was called O-strategy (modification of One-shape dimension).
(c) Producing a defective series of irregular figures. This was called N-strategy.

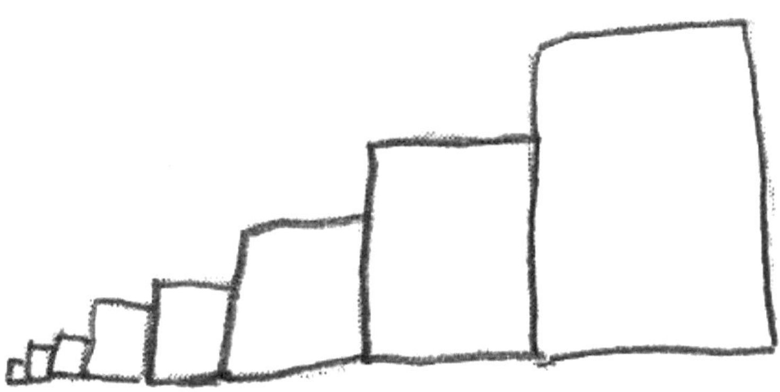

Fig. 5.1 Increasing both dimensions (T-strategy), from Gagatsis, Sriraman, Elia, & Modestou (2006, p. 34), with permission from Nordic Studies in Mathematics Education

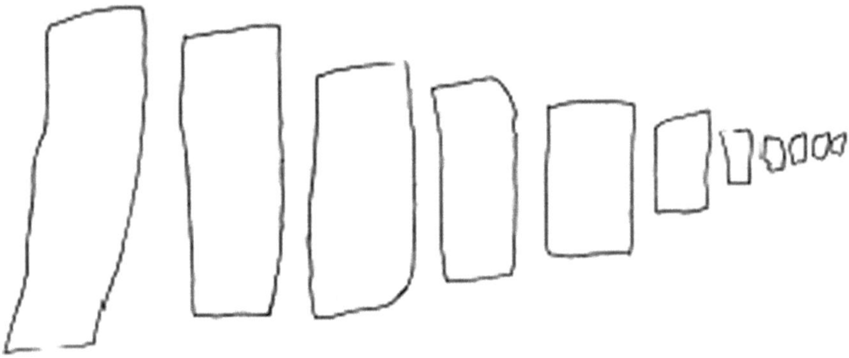

Fig. 5.2 Decreasing both dimensions (T-strategy), from Gagatsis et al. (2006, p. 34), with permission from Nordic Studies in Mathematics Education

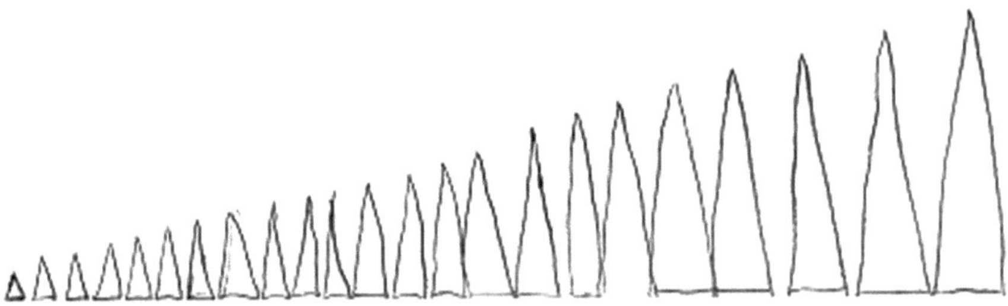

Fig. 5.3 Increasing only one dimension (O-strategy), from Gagatsis et al. (2006, p. 35), with permission from Nordic Studies in Mathematics Education

Fig. 5.4 Squares in the series of rectangles, from Gagatsis et al. (2006, p. 35), with permission from Nordic Studies in Mathematics Education

Fig. 5.5 Rectangles in the series of squares, from Gagatsis et al. (2006, p. 35), with permission from Nordic Studies in Mathematics Education

The findings of the study and specifically the results of the hierarchical similarity analysis (Gras, Suzuki, Guillet, & Spagnolo, 2008) showed that the children tended to use each strategy in a consistent way across the different figures and types of optic modification (increasing or decreasing size) in their attempt to solve the tasks. This kind of behavior may be attributed to children's different ways of looking at geometrical figures.

The application of O-strategy by many children to different tasks signifies a global apprehension of the figure with partial analysis and use of its figural units (Duval, 2014). Children following the O-strategy were not confined to keeping the prototypical form of the initial shape of the pathway. They were more inclined on the one hand to focus on the attributes they regarded as critical for the geometrical figure and keep them unchanged (e.g., number of sides, parallel or vertical sides), and on the other hand, to give less attention to the attributes, they considered as non-critical and alter them (e.g., ratio of dimensions). While in the paths of rectangle and triangle this alteration did not lead to a change in the type of figures (they remained rectangles or triangles), in the path of squares this alteration resulted in a change into (non-square) rectangles. This deficit might stem from children's lack of understanding of the relationship between squares and rectangles, a possible consequence of either prior teaching (Sarama & Clements, 2009) or not being taught geometrical shapes. Nevertheless, the capacity that these children demonstrated in visualizing dynamically the optic transformation of geometrical figures cannot be ignored. O-strategy can be seen as an indication of children's initial steps in the development of the understanding that a geometrical figure is a representation of an object and not the perceptual object itself which has to keep its initial form.

The use of T-strategy may be a result of children's attempt not to change the holistic appearance of the figure. It is likely that in dealing with the optic transformation tasks these children mainly used visual perception. This is a simple way of looking at a figure as for any image of material objects, which hinders the recognition of the geometrical objects represented (object recognition) (Duval, 2014). A further explanation, related to the former one, is the possibility that prototypical images of shapes were central to children's reasoning. These limited views of the particular shapes may have reinforced children's tendency to keep the prototypical form of the initial shape constant in the series they produced, because, otherwise, different forms of shapes may have occurred, not as "good" as the first one (Mesquita, 1998).

The construction of a defective series of irregular figures (N-approach), which was also applied consistently by the children, suggests that these children had not yet developed the ability to discriminate visually between shapes and could not even focus on the visual form of a figure irrespective of changes in size. Furthermore, it is possible that the children encountered difficulties in interpreting the verbal information in the task instructions as visual terms needed to form images for the geometrical figures and their optic modification.

Children's approaches toward the tasks varied with respect to their age. O-strategy was used mostly by seven- to eight-year-old children, while N-approach was used primarily by four- to six-year-old children. T-strategy was used mostly by children of six to seven years of age in some figures and four- to six-year-old children in other figures. Children's age, combined with the strategy they used, offered further support to the above interpretations of both children's ways and difficulties in looking at and making sense of geometrical figures.

Children's ability to recognize triangles, squares, and rectangles was found to connect only marginally with their responses to the transformation tasks. Thus, it can be concluded that shape identification requires different types of abilities from shape construction and transformation, processes that involve dynamic visualization processes.

5.6 Preschoolers' Imaginary Perspective Taking

The perception of spatial relationships, that is, whether a person can relate spatially several objects to one another or to oneself, is a major visualization and spatial reasoning ability (Gutiérrez, 1996). This ability is pertinent to mentally taking a particular point of view, an important aspect of spatial visualization that is often found in children's everyday activities (e.g., game "hide-and-seek"). To further explore the development of spatial visualization abilities in early childhood, Van den Heuvel-Panhuizen, Elia and Robitzsch (2015) set up a study focusing on the performance of preschoolers in mentally representing a viewpoint different from one's own, namely "imaginary perspective taking" (IPT). A major concern of the study was also to find out whether there were cross-cultural patterns in the IPT performances by children in different countries.

The ability of IPT can be divided into subcomponents. Flavell, Abrahams, Croft, & Flavell (1981) proposed and validated a distinction for these subcomponents into two abilities of perspective taking. Both were tested through tasks with cards placed between the experimenter and the child, in which the child had to take the experimenter's perspective. The so-called Level 1 competence concerns the visibility of objects. It implies the ability to deduce which objects are visible or not from the other viewpoint. To determine whether an object is visible, a possible strategy is to imagine oneself in the other position, projecting an observer's line of sight and verifying whether the target object meets with this line (Yaniv & Shatz, 1990). However, Michelon and Zacks (2006) suggested that this Level 1 competence might also require a line-of-sight tracing, that is, an imaginary process that acts as if an actual line is drawn between the other observer and the target object. Thus, while the former strategy in Level 1 competence involves the use of a dynamic mental image, the latter strategy is based on the creation of a concrete image (Presmeg, 1986).

The Level 2 competence concerns the appearance of object. It implies the ability to indicate how an object looks when it is seen from a different viewpoint. This competence requires a child to deal with multiple aspects of the visual appearance of an object, including features such as size, shape, and location, and to understand that these features differ when an object is seen from different perspectives

(Pillow & Flavell, 1986). Level 2 competence therefore principally involves the use of dynamic mental imagery (Presmeg, 1986), as it requires applying specific knowledge about how changes in the observer–object relationship influence aspects of the appearance. Flavell et al. (1981) found that both IPT competences are acquired by children as young as five years of age. Specifically, three-year-olds performed well on Level 1 tasks but had difficulties with Level 2 tasks, even after a brief training. Usually, this Level 2 competence is attained at about four or five years of age (Pillow & Flavell, 1986).

In Van den Heuvel-Panhuizen, Elia, & Robitzsch's study (2015), a survey was carried out in the Netherlands and in Cyprus. Table 5.1 shows the sample composition of the preschoolers that participated in the study. Their performance in IPT was assessed by administering two test booklets, each with pictorial paper-and-pencil items about imagining visibility (IPT type 1) and about imagining appearance (IPT type 2). For example, the Tower item (Fig. 5.6) is used to measure IPT type 1. In this item, the children were asked what the girl who stands on top of the tower sees. The Mouse item (Fig. 5.7) measures IPT type 2. Here, the children had to determine how the mouse looks from above.

The study confirmed previous studies' findings (e.g., Flavell, Abrahams Everett, Croft, & Flavell, 1981) that development of the IPT type 1 competence (visibility) precedes the IPT type 2 competence (appearance). Preschoolers in the Netherlands and Cyprus answered on average, respectively, 70 and 55% of the visibility items correctly, 40 and 30% of the appearance items correctly. For the visibility items, these percentages are more or less in agreement with Flavell et al. (1981), but not for the appearance items. These findings are confined by the nature of the study's items, which included drawings representing the objects, and the environment in which the objects (and sometimes also the observer) were situated (2D representations). In Flavell's studies, the perspective taking tasks were situated in concrete situations, mostly with physical objects (3D displays). These findings imply that the creation and use of dynamic mental imagery with high cognitive demands in IPT

Table 5.1 Sample composition

Child characteristic	Group	Number of children			
		NL (*N* = 334)		Cyprus (*N* = 304)	
Preschool year	K1	123		86	
	K2	211		218	
Gender	Boys	176		141	
	Girls	158		163	
Age		Age			
		M	SD	*M*	SD
	K1	4.67	0.38	4.67	0.28
	K2	5.69	0.37	5.61	0.32
	K1 + K2	5.32	0.62	5.35	0.53

Fig. 5.6 Tower item

Fig. 5.7 Mouse item

(IPT type 2 competence) are more difficult when it is based upon pictures rather than on concrete objects. This is not the case when visualization in IPT includes processes of lower cognitive demands (IPT type 1 competence), as the children performed well in both pictorial and concrete situations.

Using regression analysis on the data showed that in both countries, the children's mathematics ability (based on scores on a test developed by the Central Institute for Test Development, *Cito*, for children in the Netherlands and on teachers' perceptions of their students' level in mathematics for children in Cyprus) was significantly positively related to IPT performance (the Netherlands: $B = 0.06$, $SE = 0.01$, $t = 7.05$, $p < 0.001$, $\eta^2 = 0.036$; Cyprus: $B = 0.04$, $SE = 0.01$, $t = 4.02$, $p < 0.001$, $\eta^2 = 0.020$). For the Netherlands' sample, the older children (in K2) significantly outperformed those in K1 ($B = 0.04$, $SE = 0.02$, $t = 2.01$, $p = 0.044$, $\eta^2 = 0.001$), while in Cyprus, the group year was not found to be a significant predictor of the IPT scores. Results also showed that in both countries there was no significant effect of gender on the IPT scores.

Although the children in the Netherlands and Cyprus may have grown up in a culturally different environment (southern versus northern Europe), the findings in the two samples generally were quite similar. In fact, the main striking difference was that in the Netherlands the children performed higher on both IPT types than those in Cyprus. An explanation could be found in the implementation of the preschool curriculum. Even if in both countries spatial reasoning is a part of the mathematics program, this does not mean that the topic is adequately implemented by the teachers of both countries. Another explanation for the performance difference might be that the children in Cyprus are less familiar than the Netherlands' children with taking a class-administered paper-and-pencil test.

5.7 Making Connections Between Space and Shape Aspects and Their Verbal Representations: The Role of Gestures

McNeill (1992) suggests that "[s]peech and gesture are elements of a single integrated process of utterance formation" (p. 35). However, in contrast to speech, which can be decomposed into parts with isolated meanings, gesture is immediate and represents an image which depends on the whole. Thus, gestures are strongly connected with visualization and can be regarded as a strong indication of this mode of thinking (see also Presmeg, 2006). This view is adopted in a study we conducted to investigate the role of gestures in manipulating and communicating spatial concepts and concepts of shape at preschool level (Elia et al., 2014). In this study, we examined a five-year-old child while interacting with her teacher in the context of a geometrical activity, based on a task which involves a major visualization ability (Gutiérrez, 1996), that is, the understanding and operating on relationships between various positions in space (Sarama & Clements, 2009). Moreover, the task required semiotic transformations, that is, conversions between spatial representations and verbal descriptions.

The activity had the form of a game for two players, one of whom was the child's teacher. During the activity, the child and the teacher sat opposite each other with a screen-divider to hide each other's work. The activity included three parts. In Part 1 and Part 3, the child freely created a construction with wooden blocks (Fig. 5.8) and then described the structure, step-by-step. The teacher built the construction using blocks from the child's verbal directions. In Part 2 of the activity, the child and the teacher switched roles, and during her verbal description, the teacher produced gestures.

In Fig. 5.8, we present the child's construction in Part 3 of the activity. The figure includes also the description of her construction and her gestural production which was then analyzed to unravel the role of gestures in using and communicating spatial and shape-related ideas by the child.

The child was found to use gestures throughout the whole part of the activity in which she acted as a describer. This finding provides evidence for the strong interrelations between geometrical thinking and gestures shown in previous studies. Gesture production provided support to internal spatial visualization (Chu & Kita, 2011) of various geometry aspects that the child had to describe at the same time, such as shape, size, location, and orientation of blocks.

The analysis of the child's description indicated that various space and shape aspects were visualized through gestures. Moreover, different aspects of geometrical content were more likely to stimulate the use of specific types of gestures by the child based on McNeill's (1992) classification. Specifically, when the child described the shape of a block, e.g., a cylinder (named as circle by the child, lines 31–33, see Fig. 5.9), the orientation of a block, e.g., horizontal direction of a parallelepiped (lines 29–30, see Fig. 5.10) and topological relations of proximity or separation, e.g., shapes that were attached or not (lines 6–8, see Fig. 5.11), she tended to produce iconic gestures (McNeill, 1992) which depicted the geometry aspects involved. Interestingly, in some cases when explaining the placement of blocks, the child produced iconic gestures to represent mental images of kinesthetic character (Presmeg, 1986). Specifically, these iconic gestures visualized the movement of placing the objects in their current location (lines 11–12, see Fig. 5.12). In other cases, when the child explained the location of the blocks in her construction (e.g., in front), she used deictic gestures (McNeill, 1992), indicating the position in which the blocks were placed (lines 26–27, see Fig. 5.13). In summary, iconic gestures served multiple functions in the child's geometrical thinking and were used more often relatively to deictic gestures which were rather mono-functional.

The child's gestures were found not only to represent visually the geometrical information (e.g., naming of shapes) given by her verbal expressions, thus reinforcing the meaning of speech (e.g., lines 31–33, Fig. 5.9), but also to visually complement, enrich, and specify her verbal descriptions, particularly when her verbal utterances were unclear, general, or incomplete. Some space and shape

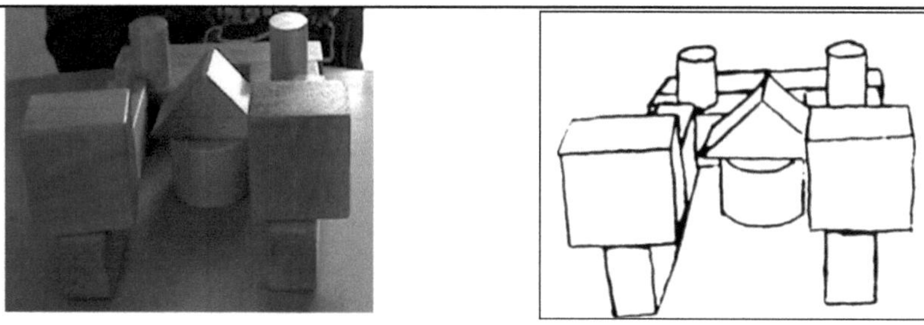

Child: Take two long shapes. [Stretches out one hand vertically to her body and forms a straight line in the air by moving her hand with the palm open near her chest].
Two.

Teacher: How shall I put them?

Child: Like this. [Moves her hands away from one another with one palm opposite the other]
Not like this. [Puts the palms of her hands together]
Take two squares.

Teacher: Where shall I put them?

Child: On [Moves her hand downwards pretending to hold the block and place it on another block] the ….. [Stretches out her hands vertically to her body and forms a straight line in the air by moving her hands with the pointing fingers stretched near her chest] but not like this [takes a parallelepiped with dimensions of similar length and turns it widthwise], like this [turns the parallelepiped lengthwise].
Then take two [Shows two and then three fingers] small … [Moves her pointing finger to form a semicircle], two small [Moves her pointing finger to form a semicircle], two small bridges.
Then put them attached [Joins the fingers of her hands] to the long shapes [Stretches out one hand vertically to her body and forms a straight line in the air by moving her hand with the palm open near her chest], in front of them, not here [Points with both her hands behind the parallelepipeds], but here [points with both her hands in front of the parallelepipeds, close to her].
Then take another long shape.
Put it in front of the bridges [Opens her hands to form a flat surface and joins her fingers in front of the bridges, close to her].
Take two circles [Makes a round line vertically in the air with her pointing finger], but small ones [Moves her hands close to her face and forms fists].
Put them on the bridges [Moves both her hands downwards pretending to hold the blocks and place them on other blocks].

Fig. 5.8 Construction made by the child, in Part 3 of the activity, followed by the child's description of her construction, from Elia et al. (2014, p. 745)

Fig. 5.9 Iconic gesture for the shape of cylinder (lines 31–33), from Elia et al. (2014, p. 745)

aspects of the construction were manifested mainly by the child's gestures instead of her words. For example, the orientation of a shape on the plane was a spatial aspect that was never expressed verbally by the child during the activity, but started to become explicit thanks to the iconic gestures she produced. Even when the child verbally explained the location of a block (e.g., in front of another block), she simultaneously used a gesture to illustrate the orientation of that block, that is, whether it was horizontally or vertically positioned (lines 29–30, see Fig. 5.10). These iconic gestures seemed to be essential and valuable in visually representing the orientation of a shape and other geometrical concepts that were complex for the child. It can be claimed that the child's visualization through gestures, together with her verbalization through speech, were harmonically coordinated to successfully accomplish the given description task.

The child was found to take the teacher's gestures as a visual model in describing her construction after observing the teacher's corresponding verbal and gestural description. In one case, she mimicked and extended the teacher's gesture

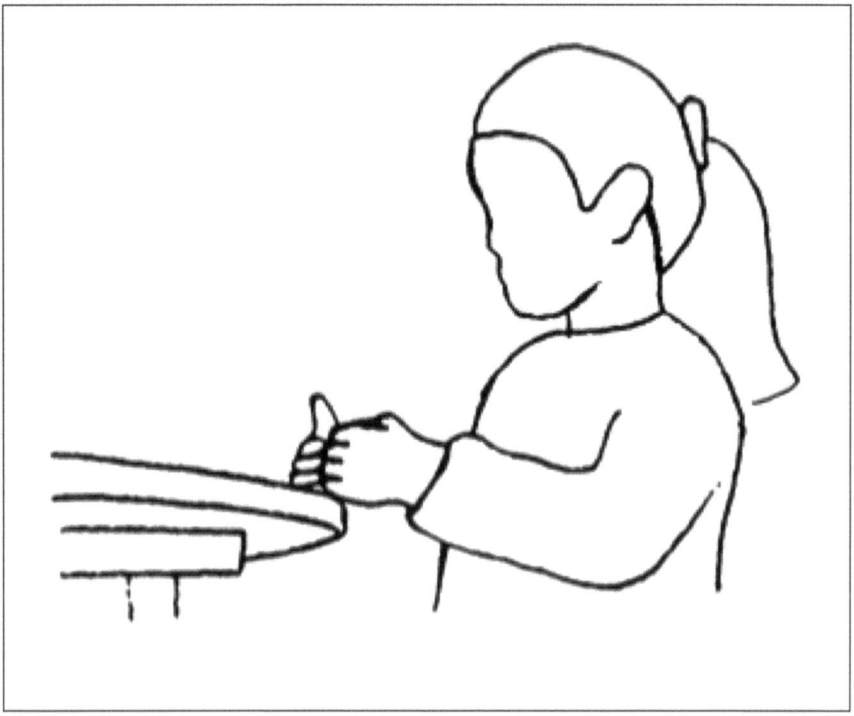

Fig. 5.10 Child's iconic gesture for the orientation of the parallelepiped in front of the blocks named "bridges" (lines 29–30), from Elia et al. (2014, p. 747)

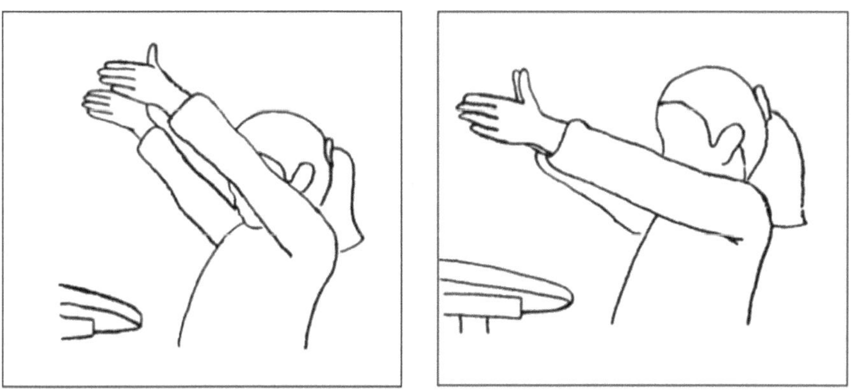

Fig. 5.11 Child's gesture supplementing the verbal expressions "like this" (left) and "not like this" (right) about the positions of two parallelepipeds (lines 6–8), from Elia et al. (2014, p. 751)

Fig. 5.12 Child's iconic gesture to represent the movement of placing the block on another block (lines 11–12)

that depicted the spatial relation of proximity between two blocks (i.e., placing the palms of her hands together) by using the same gesture and also by adding a contrast to it in a similar situation. Prior to the teacher's description, the child did not make any reference (either verbal or visual) to the proximity between blocks. Specifically, after the teacher's description, the child used a gesture that represented the relative position (separation) of two blocks in her construction, and then a gesture to show how this spatial relation was opposed to the image of two attached blocks (counterexample) (lines 6–8, see Fig. 5.11), that had been represented previously by the teacher's gesture. This change in the child's verbal and gestural acts provides evidence for the contribution of the teacher's expression of the spatial relation of proximity through gestural and speech production on the child's visualization of the particular concept. Furthermore, the contrast added by the child indicates that she internalized and creatively used the meaning of the gesture she observed and then produced visually in her own description. This finding indicates the positive influence on the child's learning of these concepts from the teacher's gestures and verbal expressions representing her mental images of spatial concepts.

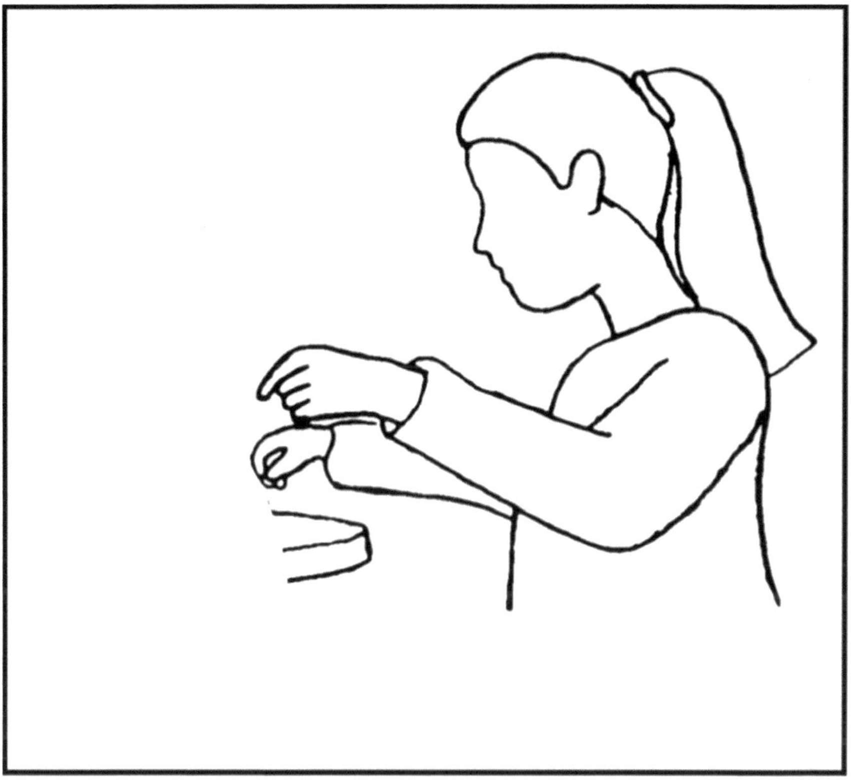

Fig. 5.13 Child's deictic gesture for the position of two "bridges" in front of the two parallelepipeds (lines 26–27)

5.8 Concluding Remarks

The work described above investigated geometry learning in the early years in three different aspects of the development of understanding of shapes and space: transformation of 2D geometrical figures, IPT, and spatial concepts and 3D shapes. Despite the different approaches that were used in this work, all three studies provided evidence for the essential role of visualization in the development of geometrical thinking. This work also revealed how young children use visualization in various types of geometrical activity including either plane or spatial geometry, indicating the multiple qualities and uses of visualization and visual skills in early geometry learning.

Based on the above, it is suggested that geometry learning in the early years cannot be examined without taking into account children's visual reasoning. However, the investigation of visual imagery is a rather difficult task, as a researcher cannot be sure that his/her interpretation about what another person has in his/her mind is accurate (Presmeg, 2014). This investigation is even more challenging in research which focuses on early childhood, as children are still developing their mathematics communication abilities (Van Oers, 2013). Nevertheless, the tasks that were used in the three studies, open dynamic transformation tasks of geometrical figures in the first study, paper-and-pencil pictorial perspective taking tasks with 2D representations of real situations in the second study, and a task involving a child–teacher–child interaction while building a spatial construction and describing it, were found to be successful in this endeavor. They enabled us to identify the use of visualization in the ways children were making sense of space and shape concepts, by means of their own dynamic drawings, selection of given illustrations, or gestures and speech. This shows that selecting or developing appropriate tasks which evoke children's geometrical thinking in meaningful ways, not confined to the simple perceptual apprehension of figures but to the stimulation of deeper spatial insights, is crucial in investigating (and also in developing) visualization in early geometry learning.

This focus on children's visualization processes was found to make "visible" implicit aspects of children's geometrical reasoning. Thus, finding ways to support children in making explicit their visual images and processes in geometry could be useful in assessing children's geometrical thinking. For example, the various strategies children use in dynamic transformation tasks of geometrical figures may enable the teacher to recognize whether children at an early age conceptualize geometrical figures as representations of objects or simple drawings and thus provide the appropriate support. Moreover, giving attention both to the verbal behavior and to the gestural production of children, which reveals the mental images they have constructed for geometrical concepts, in whole classroom interactions, in peer interactions, and in teacher–child interactions, allows the teachers to gain a better understanding of children's learning processes and outcomes and thus make their teaching better match the children's needs.

Our research findings provide evidence for the diversity of factors that may intervene in geometry learning which involves the use of visualization. Task-related characteristics is a major category of these factors, which include, for example, the competence level required by the task, the type of representation (2D or 3D), the type of geometrical figures that are involved, and other cognitive demands. Another category of such factors refer to children's characteristics, including, for example, children's mathematics ability, age, preschool year, and culture. Finally, our work indicates that a number of factors related to teachers and teaching might have an influential role in young children's geometry learning and visualization of space and shapes. These include teachers' gestural production and speech in their interactions with the children, with the teachers' iconic gestures of geometry aspects and corresponding words having the strongest influence on children, as well as the emphasis given by the teachers on spatial reasoning in the teaching of geometry.

The role of the teacher in early geometry learning and in the development of visual reasoning in this domain and determining what is necessary to fulfill this role is an issue that needs further research. Specifically, future studies with more children, longer observations, and a variety of geometrical problem-solving tasks need to be conducted before deriving the characteristics of teaching from a gestural and verbal perspective which can be beneficial for learning in early geometry and the development of visualization abilities. Furthermore, future studies should include in-depth analyses of possible differences in the educational and cultural environment of children which might be associated with the considerable variability found in IPT in children. It would be worthwhile also to explore systematically how the type of presentation of geometrical material to children, including 3D situations, 2D representations (such as work sheets or drawings in picture books), and conversions between 2D representations and 3D situations, influences children's IPT performance. Moreover, it would be theoretically interesting and practically important to investigate the impact of teaching that encourages "visual dynamic intuition" in transformation tasks of geometrical figures on young children's understanding of shape properties and characteristics, as well as of the interconnectivity and hierarchical commonalities and differences among shapes, such as rectangles and squares. Finally, it would be interesting to study the three visualization approaches described in this chapter further, with a particular focus on how they can be connected to contribute to the development of geometrical concepts in early childhood.

References

Alibali, M. (2005). Gesture in spatial cognition: expressing, communicating, and thinking about spatial information. *Spatial Cognition and Computation, 5*(4), 307–331.

Arcavi, A. (2003). The role of visual representations in the learning of mathematics. *Educational Studies in Mathematics, 52*(3), 215–241.

Bishop, A. J. (1983). Space and geometry. In R. Lesh & M. Landau (Eds.), *Acquisition of Mathematics Concepts and Processes* (pp. 175–203). New York: Academic Press.

Castelnuovo, E. (1972). *Documenti di un' esposizione de Matematica.* Torino: Boringhieri.

Chu, M., & Kita, S. (2011). The nature of gestures' beneficial role in spatial problem solving. *Journal of Experimental Psychology: General, 140*(1), 102–116.

Dindyal, J. (2015). Geometry in the early years: a commentary. *ZDM Mathematics Education, 47* (3), 519–529.

Duval, R. (1995). Geometrical Pictures: Kinds of representation and specific processes. In R. Sutherland & J. Mason (Eds.), *Exploiting mental imagery with computers in mathematical education* (pp. 142–157). Berlin: Springer.

Duval, R. (2014). The first crucial point in geometry learning: visualization. *Mediterranean Journal for Research in Mathematics education, 13,* 1–28.

Egsgard, J. C. (1970). Some ideas in Geometry that can be taught from K – 6. *Educational Studies in Mathematics, 2*(4), 478–495.

Elia, I., Gagatsis, A., & Van den Heuvel-Panhuizen, M. (2014). The role of gestures in making connections between spatial and verbal representations in the early years: Findings from a case study. *Mathematics Education Research Journal, 26,* 735–761.

Flavell, J. H., Abrahams Everett, B., Croft, K., & Flavell, E. R. (1981). Young children's knowledge about visual perception: Further evidence for the Level 1-Level 2 distinction. *Developmental Psychology, 17*(1), 99–103.

Freudenthal, H. (1973). *Mathematics as an educational task.* Dordrecht, the Netherlands: D. Reidel Publishing Company.

Gagatsis, A., Sriraman, B., Elia, I., & Modestou, M. (2006). Exploring young children's geometrical strategies. *Nordic Studies in Mathematics Education, 11*(2), 23–50.

Gras, R., Suzuki, E., Guillet, F., & Spagnolo, F. (Eds.). (2008). *Statistical implicative analysis: Theory and applications*. Heidelberg: Springer.

Gutiérrez, A. (1996). Visualization in 3-dimensional geometry: In search of a framework. In L. Puig & A. Guttierez (Eds.), *Proceedings of the 20th conference of the International Group for the Psychology of Mathematics Education* (vol. 1, pp. 3–19). Valencia: Universidad de Valencia.

Hoffer, A. (1981). Geometry is more than proof. *Mathematics Teacher, 74,* 11–18.

Krauss, R. M., Chen, Y., & Gottesman, R. (2000). Lexical gestures and lexical access: A process model. In D. McNeill (Ed.), *Language and gesture* (pp. 261–283). Cambridge, UK: Cambridge University Press.

McNeill, D. (1992). *Hand and mind: What gestures reveal about thought*. Chicago: The University of Chicago Press.

Mesquita, A. (1998). On conceptual obstacles linked with external representation in geometry. *Journal of Mathematical Behavior, 17*(2), 183–195.

Michelon, P., & Zacks, J. M. (2006). Two kinds of visual perspective taking. *Perception and Psychophysics, 68*(2), 327–337.

Moss, J., Hawes, Z., Naqvi, S., & Caswell, B. (2015). Adapting Japanese Lesson Study to enhance the teaching and learning of geometry and spatial reasoning in early years classrooms: a case study. *ZDM Mathematics Education, 47*(3), 377–390.

National Council of Teachers of Mathematics (NCTM). (2000). *Principles and standards for school mathematics*. Reston, VA: Author.

Nemirovsky, R., & Ferrara, F. (2009). Mathematical imagination and embodied cognition. *Educational Studies in Mathematics, 70,* 159–174.

Pillow, B., & Flavell, J. (1986). Young children's knowledge about visual perception: Projective size and shape. *Child Development, 57,* 125–135.

Presmeg, N. (2014). Contemplating visualization as an epistemological learning tool in mathematics. *ZDM Mathematics Education, 46,* 151–157.

Presmeg, N. C. (1986). Visualization in high school mathematics. *For the Learning of Mathematics, 6,* 71–81.

Presmeg, N. C. (2006). Research on visualization in learning and teaching mathematics: Emergence from psychology. In A. Gutiérrez & P. Boero (Eds.), *Handbook of research on the psychology of mathematics education* (pp. 205–235). Rotterdam: Sense Publishers.

Sarama, J., & Clements, D. H. (2009). *Early childhood mathematics education research: learning trajectories for young children*. New York: Routledge.

Sinclair, N., & Bruce, C. (2015). New opportunities in geometry education at the primary school. *ZDM Mathematics Education, 47*(3), 319–329.

Soury-Lavergne, S., & Maschietto, M. (2015). Articulation of spatial and geometrical knowledge in problem solving with technology at primary school. *ZDM Mathematics Education, 47*(3), 435–449.

Sutherland, R. (1995). Mediating mathematical action. In R. Sutherland & J. Mason (Eds.), *Exploiting Mental imagery with Computers in Mathematics Education* (pp. 71–81). Berlin: Springer-Verlag.

Van den Heuvel-Panhuizen, M. & Buys, K. (Eds.). (2008). *Young children learn measurement and geometry.* Rotterdam, the Netherlands: Sense Publishers.

Van den Heuvel-Panhuizen, M., Elia, I., & Robitzsch, A. (2015). Kindergartner's performance in two types of imaginary perspective taking. *ZDM Mathematics Education, 47*(3), 345–362.

Van Hiele, P. M. (1985). *The child's thought and geometry.* In D. Geddes & R. Tischler (Eds.), English translation of selected writings of Dina van Hiele-Geldof and Pierre M. van Hiele (pp. 243–252). Brooklyn: Brooklyn College, School of Education (Original work published 1959).

Van Oers, B. (2013). Communicating about number: fostering young children's mathematical orientation in the world. In L. English & J. Mulligan (Eds.), *Advances in mathematics education: reconceptualizing early mathematics learning* (pp. 183–203). New York: Springer.

Yakimanskaya, I. S. (1991). *The development of spatial thinking in school children.* (Soviet Studies in Mathematics Education, vol. 3). Reston, USA: NCTM.
Yaniv, I., & Shatz, M. (1990). Heuristics of reasoning and analogy in children's visual perspective taking. *Child Development, 61,* 1491–1501.

Authors Biography

Dr. Iliada Elia is an Assistant Professor of Mathematics Pedagogy in Early Childhood at the Department of Education of the University of Cyprus. In the years 2009–2010 and 2012–2015, she was a guest researcher at the Freudenthal Institute of Utrecht University. Her research interests lie in the field of mathematics education with a focus in the early years and include among others the semiotic approach in the learning of mathematics, the evolution of geometry thinking, and arithmetic problem solving. She has been involved in national and international projects on various topics in the research field of mathematics education.

Prof. Dr. Marja van den Heuvel-Panhuizen is a Full Professor of Mathematics Education at Utrecht University and is affiliated both with the Freudenthal Institute of the Science Faculty and with the Freudenthal Group of the Department of Education and Pedagogy of the Faculty of Social and Behavioral Sciences. Since 2016, she is Emeritus Professor, but her work continues. She started to work at the Freudenthal Institute in 1987 and was a Visiting Professor at Dortmund University and at IQB, Humboldt University Berlin. She was and is involved in many national and international projects in primary education, special education, and early childhood.

Prof. Athanasios Gagatsis is a Full professor of Mathematics Education at the Department of Education of the University of Cyprus. Athanasios is the Vice Rector of Academic Affairs of the University of Cyprus since December 2010. He was Lecturer and Assistant Professor at the Department of Mathematics of the Aristotle University of Thessaloniki and Visiting Professor and Associate fellow in different European Universities. His research is published in different languages on geometry teaching and learning, on the use of representations in the teaching and learning of different mathematical concepts, on the history of mathematics education, and on implicative statistical analysis.

Chapter 6
A Possible Learning Trajectory for Young Children's Experiences of the Evolution of the Base-10 Positional Numeral System

Mun Yee Lai and Chun Ip Fung

Abstract The concept of base-10 numeral system is fundamental to other mathematical concepts such as decimal numbers and exponents, and yet for a variety of reasons, including differences in number naming language, many children in the early years of school find the concept of place value difficult. Renowned mathematics educator Freudenthal (Mathematics as an educational task. Reidel, Dordrecht, Holland, 1973) was critical of the way that mathematical content knowledge is often simplistically identified as the learning of ready-made mathematical objects, leaving the evolution and refinement process of mathematics unattended. It seems, therefore, that children's experience of the evolution and construction of the base-10 numeral system should gain a position in young children's mathematical experiences. This chapter outlines a possible learning trajectory for children in the early years of school for this purpose, together with a brief elaboration of its theoretical underpinnings. The ultimate goal is to open up possibility for young children to connect to the evolution of the base-10 number system, and so deepen their understanding of the concept of place value.

Keywords Base-10 positional numeral system · Concept of place value
Teaching for mathematizing · Early years of school

M. Y. Lai (✉)
Australian Catholic University, Melbourne, Australia
e-mail: MunYee.Lai@acu.edu.au

C. I. Fung
The Education University of Hong Kong, Hong Kong, China
e-mail: cifung@ied.edu.hk

V. Kinnear et al. (eds.), *Forging Connections in Early Mathematics Teaching and Learning*, Early Mathematics Learning and Development,
https://doi.org/10.1007/978-981-10-7153-9_6

6.1 Introduction

Number and counting in early childhood mathematics are two connected concepts (Haylock & Cockburn, 2008). A child's ability to count by reciting number names does not, however, mean that the concept of 'number' has been established, as the counting strategy can be purely routine and procedural. Munn (2008) has reminded teachers that young children who can 'count' fluently are reciting number names by rote and that this can be done without necessarily understanding number. We know also that learning number sequences fluently can be problematic for young children, with common counting errors for teen numbers and decade transitions, for example moving from 29 to 30 (Geary, 2000). One influence on these errors is language, as unlike number words in Asian languages, in European languages such as English, two-digit number words can be quite arbitrary (e.g. for numbers between 10 and 15) and do not correspond with any elegance to the base-10 positional structure. When children, for example, say *eleven* or *twelve* there is no pattern, and teen numbers use language associated with ones followed by tens (e.g. *thirteen, fourteen*) and *thirty* and *thirty-one* provides little indication as to how many tens it is referring to. By contrast in the Chinese number system, numbers beyond ten are generated by consistent and logical rules which follow and embody the notion of the base-10 system of Hindu-Arabic numerals (Han & Ginsburg, 2001; Ng & Rao, 2010). For example, 11 is *shi yi* (*shi* = one ten and *yi* = one) and so states how many tens and how many ones and the place value order. Similarly, 12 is *shi ér* (one ten and two), 30 is *san shi* or three tens, and 31 is *san shi yi* (three tens and one). The transparency of place value in the number system may account for the apparent ease for learning numbers past 10 (Ng & Rao, 2010).

The lack of transparent correspondence between English two-digit number words and the base-10 positional system of the Hindu-Arabic numerals can lead English-speaking children to understand two-digit numbers words as referring to counted collections of single objects (Sharma, 1993). The disadvantages that English number language presents can amplify difficulties children have in constructing the concepts of quantity, place value and part-whole relationships for whole numbers (i.e., the relationship between the digits and the number they represent) (Camos, 2008; Moeller et al., 2011). The numeral word system in English can help explain why many children find counting after eleven or why they may miss the ten in a sequence, for example counting 38, 39 and then 41 (Ryan & Williams, 2007).

Different studies (e.g. Geary, 2000; Lawton & Hansen, 2011; Ross, 1989) have consistently reported that many young children find concepts of place value difficult and consequently commit different types of counting errors and arithmetic errors. For instance, many young children fail to differentiate between the face value of each digit in a number and the complete value of the same number (Varelas & Becker, 1997). They may understand that twenty-four or 24 represents 'how many' for a collection of 24 objects, but do not know that the digit on the left represents two sets of 10 and the digit on the right represents 4 sets of 1 (Geary, 2000; Ross, 1989). Such difficulties reflect that they have not mastered the concept of place value. Some children misread numbers like 15 and 51 or 82 and 28 because they understand them as identical in meaning, regardless of the position of digits (Sharma, 1993). Children in the early years of school have difficulty in solving basic arithmetic problems which involve carrying or borrowing from one number column to the next (Geary, 2000).

Cumulating studies such as Chan, Au and Tang (2014), Chan and Ho (2010), Collet (2003), Fuson and Briars (1990) and Ho and Cheng (1997) have indicated that mastery of the place value structure of the Hindu-Arabic numeral system is crucial to whole number comparison, arithmetic learning and problem-solving. Mathematics educators such as Sharma (1993) have pointed out that the concept of place value is one of the most crucial arithmetical concepts to be learned by early age children because it is fundamental to understanding all arithmetical algorithms. Sharma has further elaborated the importance of place value in mathematics learning:

> The mastery of the place value system is an important milestone in a child's journey on the way to the acquisition of other mathematical concepts. A child's facility with place value helps him[sic] in understanding other mathematical concepts: divisibility principles, prime numbers, scientific notation, exponents, and of course mathematical algorithms. In other words, any concept that is dependent on number is dependent on place value. (p. 2)

Although many studies have investigated the development of students' understanding of place value in primary grades, not much attention has been given to investigating preschool children's development of this concept (McGuire & Kinzie, 2013). The lack of attention may stem from concerns about the appropriateness of introducing the concept of place value in preschool (McGuire & Kinzie, 2013). For instance, not much time is spent on activities which require preschool children to group ones into tens whereas the conceptual aspect of place value in numerous studies has been focused on understanding quantities through grouping and regrouping base-10 manipulatives. Valeras and Becker (1997) have pointed out that the use of base-10 manipulatives, although appearing to represent place value quantities, does not highlight the *positional* significance of the digits in the base-10 numeral system. The base-10 manipulatives can, therefore, appear as quantities that are dissimilar to the place value system (Valeras & Becker, 1997). In order to understand place value as a concept, children must construct a positional base-10 structure and relate this conceptual structure to the digits in different positions in a multi-digit number.

This chapter aims to provide a possible learning trajectory for young children to explore and experience the evolution of a base-10 positional numeral system that fosters children's learning and understanding of a base-10 system and its connection to an understanding of the concept of place value. This is different from the study of the base-10 positional numeral system that is commonly found in the mathematics curriculum in the early years of school and is in direct contrast to the features of the base-10 manipulatives. The ultimate goal is to support children in their understanding of the place value concept in order to further their mathematics learning.

6.2 Experiencing the Evolution of the Base-10 Numeral System for Understanding the Concept of Place Value

Renowned mathematics educator Freudenthal (1973) criticized the mathematical content knowledge taught in schools as often simplistic, identifying learning as mathematical products—theorems, formulas and algorithms and separating them from their evolution and mathematical refinement processes. This separation he argued distorts both the nature and content of mathematics knowledge and distances it from the human activity that developed it. Freudenthal offered the term *mathematizing* to refer to the evolving processes in mathematics that organize unmathematical (or insufficiently mathematical) matters into a structure that is accessible and open to mathematical refinement. In short, mathematizing is defined as a developmental process during which primitive mathematical ideas evolve and move to sophisticated mathematical understanding. Mathematizing, therefore, provides a contrast to the reproductive, product-focused activity often found in mathematics teaching and learning, where mathematical formulae and operations are applied in order to find solutions to instructional tasks (Van Oers, 2014). Van Oers (2014) argues that merely reproducing mathematics education is like cutting the heart out of mathematics. Freudenthal (1991) claimed that the core work of mathematics teaching should emphasize the re-invention process in which both the final products *and* their evolution and refinement processes are highlighted. He believed that the design of all mathematics learning trajectories should be carried out in accordance with this point of view and that the design of mathematical tasks should carry a subtle mission:

> Children should repeat the learning process of mankind, not as it factually took place but rather as it would have done if people in the past had known a bit more of what we know now. (p. 48)

Building on Freudenthal's seminal work, Fung (1999) identified a learning gap (Fig. 6.1) between teaching mathematical products and their application, and teaching the process of evolution and refinement of mathematics. To fill the learning gap, Fung (1999) emphasized the importance of engaging pupils in the process of mathematizing. He proposed a framework, *Teaching for Mathematizing*, which utilizes a design science approach to provide children with the opportunity to reinvent mathematics.

During the transition from preschool to the early years of school, it is important to let children both appreciate and understand the evolution of the numeral system in order to enhance their understanding of the base-10 system and place value through by taking part in a more elaborated learning trajectory.

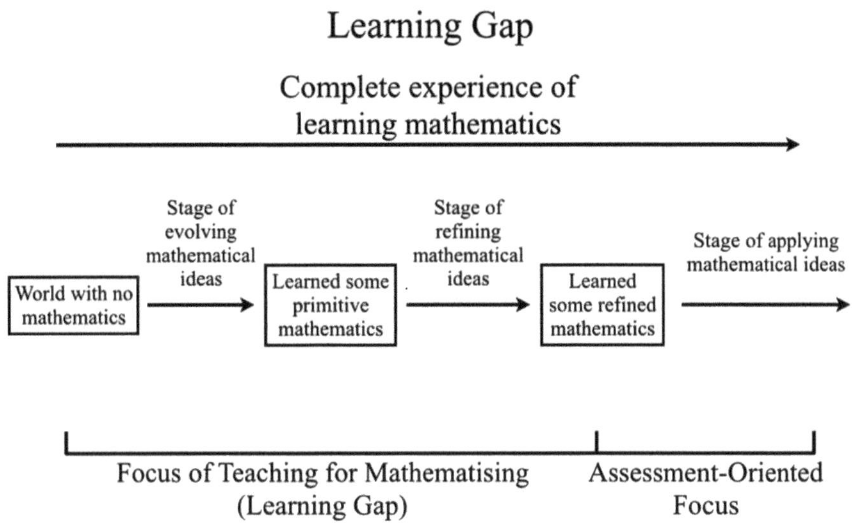

Fig. 6.1 Learning gap in mathematics education (Fung, 2004)

6.3 A Possible Trajectory for the Development of the Numeral System

Before describing a series of learning activities for children to experience the evolution of a positional numeral system, an elaboration of the mathematical underpinnings of these activities is presented.

6.3.1 The Development of the First Number Notation

The most primitive way to denote whole numbers is by repeating the same movement, action or written recording indefinitely. An historic example is the Ishango bone where one notation, a mark on the bone, represented the number '1' (one), called the first nodal number. Here we use the term 'notation' to denote a specific body movement or action (dynamic notation), a specific physical object (physical notation), a specific sound (audio notation) or a specific symbol for recording (symbolic notation). In essence, the most primitive whole number system consists of just one notation; however, such a system cannot deal with large numbers efficiently. It takes too long to act out the number using actions, or one runs out of physical objects, or the repeating sound string is too long, or the bone is not large enough for tallying. A more efficient system was needed.

6.3.2 The Development of Number Notations as Simple Grouping Systems

To improve the system, an increased number of notations was needed that corresponded numbers to notations that could be identified as quickly as possible. For instance, each of these notations; ♦ can represent one, and these notations; ♥ can represent five to denote a simple grouping system. Another example is if ‘α’ is used to denote the number notation ‘1’, ‘β’ for five times the amount represented by ‘1’ and ‘γ’ for nine times the amount represented by ‘1’. Thirteen times the amount represented by ‘1’ is then symbolized by ‘γαααα’. At this stage, the ordering of the notations is immaterial because in a simple grouping system, ‘γαααα’, ‘αγααα’, ‘ααγαα’, ‘αααγα’, ‘ααααγ’, and even the presentation in Fig. 6.2 all represent the same number.

In between the development of the most primitive and the most efficient Hindu-Arabic positional numeral system, there were various simple grouping systems used to represent numbers with different degrees of elegance. The Roman numeral system and the Egyptian hieroglyphic numeral system are well-known examples of simple grouping systems. The Roman numeral system employs a combination of letters from the Latin alphabet to signify values: ‘I’ represents one, ‘V’ represents five, ‘X’ represents ten, ‘L’ represents 50, ‘C’ represents 100, ‘D’ represents 500 and ‘M’ represents 1000. Letters are placed from left to right in order of value, starting with the largest. For example, to represent three ‘hundreds’, two ‘tens’ and one ‘one’, three Cs, two Xs and one I are needed: CCCXXI is 321. To extend the use of this system, the Roman numeral system uses both additive forms and subtractive forms. In short, a smaller number is placed after a larger number for additive forms and before a larger number for subtractive forms. For example, VI represents six and IV represents four. For bigger numbers, CX represents 110 and XC represents 90. Note that in the Roman number system, numbers are separated

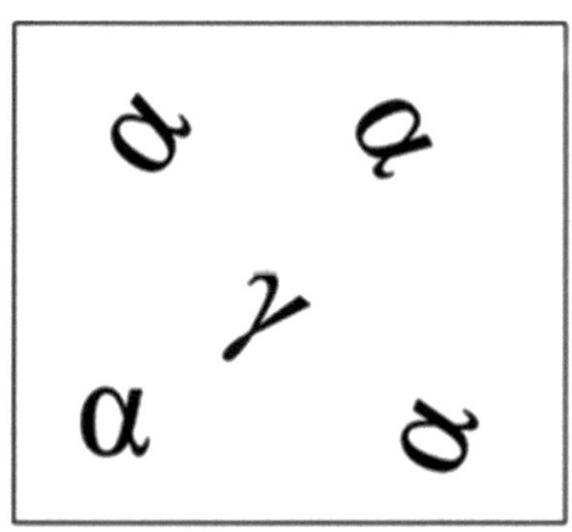

Fig. 6.2 A representation of 13 using the symbols ‘γ’ and ‘α’

into their units (i.e. into ones, tens, hundreds and thousands). Thus, 99 is 90 + 9 (XCXI) and not ‘IC’. For example, for 469, we need CD to represent 400, LX to represent 60 and IX to represent 9. Thus, CDLXIX in Roman numeral system is 469.

Simple grouping system can work for managing small numbers; however, it cannot be extended to represent all whole numbers because a finite number of notations cannot possibly correspond to an infinite number of whole numbers. In other words, we need some ‘clever’ way to apply notations so that the number of different notations needed grows at a much slower pace than the quantities being represented.

6.3.3 The Development of a Multiplicative Grouping System

The multiplicative grouping system evolved to eliminate the need to repeat notations. Thus, ‘$1\gamma 4\alpha$’ would replace ‘$\gamma\alpha\alpha\alpha\alpha$’, and ‘$2\beta 3\alpha$’ would replace ‘$\beta\beta\alpha\alpha\alpha$’. This system consists of multiplicand notations (α, β, γ, …) and multiplier notations (1, 2, 3, …), the use of which must follow a specific order. By applying order restrictions, the system enables a relatively quicker recognition of the number of each notation used.

Once the sequence of number notations (i.e. multiplicands) and multipliers is defined, the representation of a number in a multiplicative grouping system can be further refined to omit the multiplicands. This can be done by writing multipliers under specifically ordered columns. For example, if we insert corresponding multipliers of γ, β, α under respective columns, $1\gamma 4\alpha$ and $2\beta 3\alpha$ can be represented by ‘104’ and ‘023’, respectively, as shown in Figs. 6.3 and 6.4. The ‘0’ in 104, if omitted, would represent $1\beta 4\alpha$ instead of $1\gamma 4\alpha$, thus affecting the value represented. Its function is to hold a place to keep the ‘1’ on its left under the γ’s column. The use of ‘0’ in this way is as a placeholder. Writing the ‘0’ in 023, however, does not affect the value ($2\beta 3\alpha$) as represented, and ‘0’ does not act as a placeholder.

Fig. 6.3 A representation of $1\gamma 4\alpha$ using place value system

γ column	β column	α column
1	0	4

Fig. 6.4 A representation of 2β3α using place value system

γ column	β column	α column
0	2	3

6.3.4 The Development of the Base-10 Positional Numeral System

The ultimate goal for a numeral system, therefore, was to have a small collection of notations which, when appropriately combined or arranged, could be used to represent practically infinitely many whole numbers. The denary Hindu-Arabic positional numeral system makes the best use of the numerals in written form. By using just ten different notations (0, 1, 2, 3, 4, 5, 6, 7, 8, 9), called numerals, and combining them through juxtaposition according to different powers of ten (1, 10, 100, 1000, …), called place values, every whole number has an unique representation by a finite string of these numerals starting with a nonzero numeral on the left. The system uses ten as a base because of the availability of our ten fingers for counting purposes (Haylock, 2011). Any whole numbers larger than 9 are constructed using powers of ten: 10, 100, 1000, 10,000 and so on, with the direction of progression of place value, multiplying the previous value by 10, when moving from right to left. Thus, the value represented by a numeral (i.e. 0, 1, 2, 3, 4, 5, 6, 7, 8 or 9) is determined by the position of that numeral in the whole number. For instance, in the number 1234, 4 represents four ones (i.e. 1), 3 represents three tens (i.e. 10), 2 represents two hundreds (i.e. 100) and 1 represents one thousand (i.e. 1000). In the base-10 Hindu-Arabic positional numeral system, these powers of ten can continue indefinitely with higher powers. When there is no value for a given power of 10, the corresponding position is occupied by zero, the placeholder. With these principles, the base-10 Hindu-Arabic positional numeral system can represent infinitely many whole numbers.

6.4 How This Base-10 System Could Reasonably Evolve in an Early Childhood Classroom—A Possible Learning Trajectory

The Hindu-Arabic numeral system is an efficient place value system for representing numbers for calculations; however, children need to develop an understanding of how the system is structured to be able to use it. Lawton and Hansen

(2011) identify five key principles of the underlying structures that young children need to develop systematically as they move from counting to representing quantities as numerals:

(a) Numerals—there are only ten numerals in the system (0, 1, 2, 3, 4, 5, 6, 7, 8 and 9).
(b) Position—the columnar position of a numeral determines the value it represents. The numeral signifies the number of times a specific value (called the place value of the column) is counted. A numeral written in the column the place value of which is '1' is called the units digit. A numeral written in the column the place value of which is '5' is called the '5s' digit.
(c) Base-10—in the base-10 system, the place value of every column is 10 times the place value of the column on its right. Thus, place values are powers of 10 starting with $1(=10^{\circ})$ on the units column, and progressively increase to 10^1, 10^2, etc.
(d) Zero—'0' represents an empty column (zero as placeholder). It signifies that 0 times the place value of a column is counted. Any placeholder on the right of a nonzero digit cannot be omitted or else the value of the number will be incorrectly represented.
(e) Grouping and exchange—in the base-10 positional numeration system, once we accumulate 10 times the place value of a column, the limited numerals cannot denote it under the same column. We have to exchange them for 1 times the place value of its left column, leaving a 0 under the original column. Vice versa, a 1 on a column can exchange to 10 times the place value of its right column.

In the following section, we will look at ways of introducing the evolution of the base-10 numeral system with a specific focus on two key principles: Position and Base-10. As discussed earlier, many children do not master the concept of place value and thus are unable to understand what digits really mean in a multi-digit number. Afterwards, we explore a possible learning trajectory for developing the place value concept for children in the early years of school.

6.4.1 The Creation of the First Number Notation

The principle of place value is the basis of the Hindu-Arabic positional numeral system and fundamental to learning mathematics (Haylock & Cockburn, 2008). If as Freudenthal highlights, mathematics is the result of human activity (Askew, 2012), there is value in children actually experiencing the evolutionary process of the number system. To facilitate this experience, teachers can design learning experiences where children can create their own numeral system, following the footprints of ancient people who used just one notation and repeated it indefinitely to represent ever increasing numbers.

In the early childhood classroom, there are at least four ways to represent number notation as discussed earlier: making use of (1) body movement or action; (2) physical objects; (3) sound; and (4) tally recording. The most primitive way to denote all whole numbers is by repeating the same number notation—movement/action, physical objects, sound or tally recording indefinitely. It means that other numbers (such as 2, 3, 4 and so on) are represented by repeating the notation for the chosen number notation '1' accordingly. As discussed earlier, in essence, the most primitive system consists of just one notation. For instance, if a paper clip is chosen as the number notation, one paper clip represents 'one', two paper clips represent 'two', six paper clips represent 'six', fourteen paper clips represent 'fourteen' and so on. Teachers can provide children with different materials (such as paper clips, picks, counters.) to represent numbers and include making use of children's body parts such as clapping hands and tapping the floor.

To further develop children's concept of number notation, teachers could invite children to choose a body movement (e.g. winking) or a sound (e.g. a tap) as a number notation to represent quantities for a number game. In a number game, a teacher offers a number to a child who then acts out the chosen body movement or sound, so that other students can determine the initial number. For example, if tapping is chosen as the number notation, and 12 is the given number, the child has taps twelve times: <tap>, <tap>, <tap>, <tap>, <tap>, <tap>, <tap>, <tap>, <tap>, <tap>, <tap>, <tap>. With scaffolding, children will discover quickly that such a number notation system cannot deal with large numbers efficiently, as the repeating sound string or the body movement will be too long to quickly or easily work out what quantity is being represented. If physical materials are chosen as the number notation, the scale of physical materials needed for large numbers will be prohibitive. Thus, the pressing need is to develop a system in which numbers can be conveniently represented, no matter how large or small they are. Building on this idea, children are then led in the next stage to explore adding more number notations in order to reduce the number of repeating notations used to represent numbers. The following section discusses this point.

6.4.2 *The Creation of More Number Notations—An Introduction to a Simple Grouping System*

In the next stage of evolution, teachers can begin with an 'Exchange' game, to inspire children to see the need for creating number notations, and different ways to create them that reduce the number of repeating notations used to represent numbers. For instance, using big and small marbles, children decide how many small marbles (○) can be represented by one big marble (•). In this game, children explore and experiment with using more notations to represent numbers. If one big marble and one small marble represent five and one, respectively, two big marbles and three small marbles (i.e. ••○○○) represent 13 (for $5 + 5 + 1 + 1 + 1 = 13$). Another example is, if each clap and each tap represent ten and one, respectively, one clap and five taps denote 15 (for 10 + 1+ 1 + 1 + 1 + 1 = 15).

To further develop the concept of a simple grouping system for number notation, a teacher can introduce the activity 'Let my eye, nose and mouth tell you how many'. In this game, the teacher defines the number notation for different body movements, for example, each wink represents 3, each clap represents 2 and each air kiss represents 1. The teacher then gives a number between 1 and 11 to a child who is then acts out the number in any combination of winking, clapping and/or air kissing so that other children are able to determine the number. For example, if 5 is the given number, the child can wink once and clap once (i.e. <wink>, <clap>), wink once and air kiss two times (i.e. <wink>, <air kiss>, <air kiss>) or inhale once and air kiss three times (i.e. <clap>, <air kiss>, <air kiss>, <air kiss>). To reinforce the number representation concept, teachers can encourage children to represent their own numbers using their chosen number notations. For example, if a child chooses a large paper clip and a small paper clip to denote 4 and 1, respectively, she/he needs one large and three small paper clips to represent 7 (for 4 + 1 +1 + 1 = 7).

Children can be connected to the concept of a simple grouping system familiar through the monetary currency which is a genuine example of our daily use of the system. Money is non-positional because how we physically place the notes and coins will not affect the total value. It does not provide unique representation for every possible value because collections of notes with the same total value can always be used interchangeably. By adopting some guidelines for using the notes and coins, one may arrive at a unique representation for every possible value. Nevertheless, in our usual practice, such a system has a collection of notations for various values, and these notations are listed in descending order of magnitude with identical ones grouping together.

In this setting, with teachers' providing support with scaffolding and questioning, children are likely to perceive that representations do not take the form of a designated sequence and that the uniqueness of representations is not guaranteed even if we restrict the number of notations used to a minimum. For instance, 'γαααα' and 'ββααα' represent the same number if α, β and γ represent one, five and nine, respectively. Furthermore, in the simple grouping system, although the amount of different notations needed is limited, it is extremely difficult to compare the magnitude of different large numbers. This is amplified if one needs to make calculations.

6.4.3 An Introduction to the Multiplicative Grouping System

The limitations of the simple grouping system can be used to support children to about think about counting as an essential process to work out 'how many', and to consider the limitations of an unorganized sequence of number representation such as ααγαα, βαααβ or •o•oo. Children may start to appreciate the necessity of grouping unorganized notations into designated types into more efficient representations. Children can group and sequence number notations according to the values that the notations represent and see how these changes enable them to identify more quickly 'how many each of the notations there are'. For example, ααγαα, βαααβ and •○•○○ can be represented as 1γ4α, 2β3α and 2○3○ respectively. Each type of notation is preceded by the number of times it is counted. Any number is represented by a mixture of notations from the categories such as α, β, γ; • or ○ and their respective multipliers (1, 2, 3, 4, etc.). By doing so, children can begin to identify how many times each of the notations appears.

Learning activities can next take the form of giving a collection of notations in a simple grouping system and asking children to change them to a representation using a multiplicative grouping system. For example, teachers can provide students with some number representation cards that provoke the need to identify and group different number notations. This is followed by using multipliers (1, 2, 3, 4, etc.) to denote the number of times each of the notations appears. Children then work out the total quantity that each number representation card denotes. Figure 6.5 illustrates a sample of number representation cards and the expected student work.

The above activity can direct students to the need for a designated sequence—either from the smallest to biggest notations (i.e. 2✹ 4 ☾) or vice versa (i.e. 4 ✹ 2 ✹). In this system, however, the uniqueness of representation is still not addressed. In fact, the demand of uniqueness is not conspicuous if the number notations are chosen by random, as it does not scaffold gaining efficient recognition of what the representation denotes. For example, using the same set of number notations in Fig. 6.5, 11 can be represented as (a) 5 ✹ 1 ☾, (b) 4 ✹ 3 ☾, (c) 3 ✹ 5 ☾ or even 11 ☾. Young children may find the last representation more easily recognizable because the representation is straightforward and demands no grouping, and because each object represents one and so can be counted using one-to-one correspondence.

In the next activity, children choose their own number notations they think will enable them to recognize the representations they denote quickly. With teachers' prompting and scaffolding, children can be moved to identify ten as a benchmark,

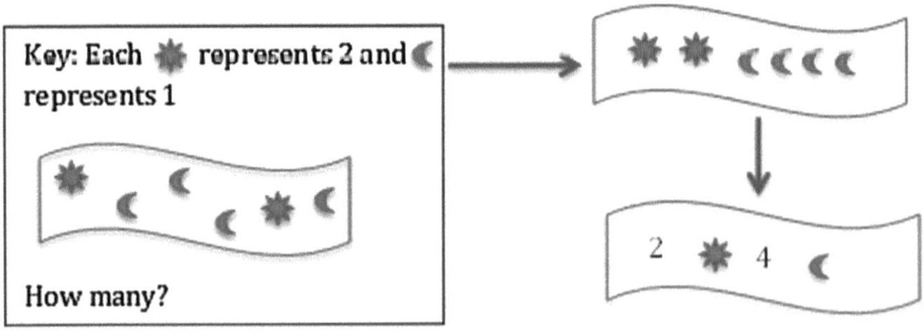

Fig. 6.5 Sample of a number representation card

with ten as a representation of ten individual objects that connect to humans having 10 fingers and 10 toes. The structural nature of ten is no different to other number words that children develop an abstract understanding of (Sharma, 1993); however, it has a specific and critical role in children making meaning of the base-10 system. Teachers can draw from children's different sets of number notations to stimulate discussion about which number notation set is better and encourage them to give reasons for their choice. Once ten as an efficient number notation is established, the uniqueness of representation from the biggest notation to the smallest (such as 2ß and 3∞ denote twenty-three for each ß and ∞ representing 10 and 1, respectively) can be derived.

The concept of ten as a representation of 10 individual objects is crucial to the conceptual development of the exchange of 10 units for a group of ten, group of ten tens into a hundred and so on for further learning the base-10 numeral system at a later stage. Activities involving grouping objects into groups of tens for recognizing the quantities quickly provide an example of this, as illustrated in Fig. 6.6.

6.4.4 *The Introduction to Base-10 Positional Numeral System—Space Organization and Place Value*

Although the concept of ten as a representation of 10 individual objects (i.e. base-10 system) and the uniqueness of representation can be developed through the previous experiences, the place value system does not emerge. A final step towards the construction of the base-10 positional numeration system is to tighten the sequence of different number notations and multipliers in a particular order, so that the unique representation for every whole number becomes visible and connects to the group organization meaning children have developed. In this section, an activity that foregrounds the spatial position of the multiplicand notations and the multipliers (i.e. digits) in a multi digit number will be explored.

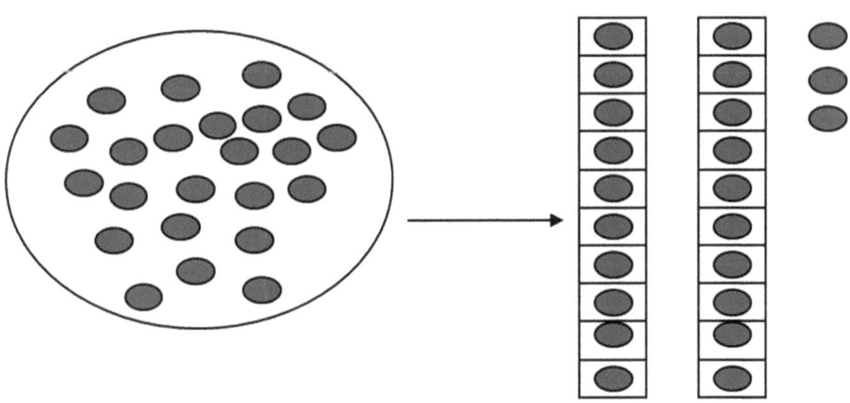

Fig. 6.6 Grouping objects into tens and ones

Valeras and Becker's (1997) research introduced the face value and complete value (FVCV) system, using pieces and board. The upper side of each of the pieces presents the face value (i.e. 1, 2, 3 …) and the underside indicates the complete value (e.g. 4 or 40 for '4' on the upper side). For example, while there are pieces with 4 on the upper side to denote a face value of 4, some of these pieces have 4 on the underside and some 40 to indicate different complete values for the same face value (refer to Fig. 6.7). The complete values are representative of the place value notation.

In this activity, children represent a quantity by placing the pieces with the correct face value and complete the value on a board on which the left column is labelled 'Tens' and the right column 'Ones' (refer to Fig. 6.8).

For instance, the quantity 46 is represented by one piece with a 6 on both the upper side and the underside placed in the right column (i.e. the 'Ones' column) on the board and another piece with 4 on the upper side but 40 on the underside placed in the left column (i.e. the 'Tens' column) on the board as shown in Fig. 6.9. In FVCV, both the position and underside of the pieces indicate the complete value represented by that piece. The spatial arrangement of the pieces conforms to the base-10 positional numeral system (Valeras & Becker, 1997).

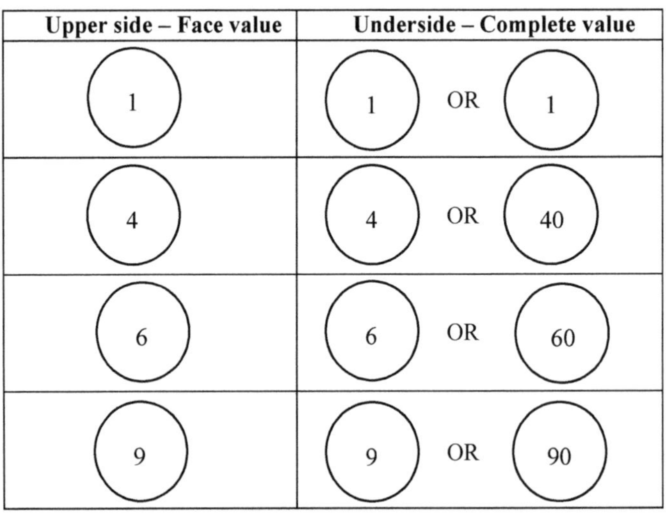

Fig. 6.7 Face value and complete value (FVCV) pieces (Modified from Valeras and Becker, 1997)

? Tens	? Ones	The number

Fig. 6.8 Board for placing face value and complete value (FVCV) pieces (Modified from Valeras and Becker, 1997)

Upper side of the pieces			Underside of the pieces	
Tens	**Ones**	**The number**	**Tens**	**Ones**
4	6	Forty-six (46)	40	6

Fig. 6.9 Pieces representing the quantity '46' on the FVCV board

The careful design of tasks and discussions can support children to connect their learning to the progression of number system development from simple groupings to understanding the structure and purpose of place value in the base-10 system, by sequencing experiences that present the fundamental reasoning behind a positional numeration system.

6.5 Conclusion

The number and structural knowledge young children need to understand the base-10 and place value systems are complex. We believe that if young children are provided with the experiences that mirror the evolution of numeration systems, they may be able to explore: (1) the effect of the choice of the multiplicands (i.e. the number notations) on the collection of multipliers; (2) the consequence of choosing the multiplicands to be a sequence of consecutive powers of certain number b (the base); (3) the differences between a multiplicative grouping system and a positional numeral system; and (4) the need of a placeholder in a positional numeral system.

Freudenthal's notion of mathematizing provides mathematics educators with a guidepost for teaching, as it emphasizes the development of carefully designed experiences that adopt an evolutionary approach to the mathematical concepts, as seen through the eyes of children as learners. These experiences need to work to meet the needs of children to reinvent the mathematics themselves and provide learning consequences that deepen children's understanding of *why* number systems are structured and organized in the way they are. In this way, the development of mathematics as a human activity and the processes at its heart are open and accessible to children, and they have an opportunity to connect to it.

References

Askew, M. (2012). Tasks. In M. Askew (Ed.), *Transforming primary mathematics: Understanding classroom tasks, tools and talk* (pp. 97–108). UK, Oxon: Routledge.

Camos, V. (2008). Low working memory capacity impedes both efficiency and learning of number transcoding in children. *Journal Experimental Child Psychology, 99*(1), 37–57.

Chan, B. M. Y., & Ho, C. S. H. (2010). The cognitive profile of Chinese children with mathematics difficulties. *Journal of Experimental Child Psychology, 107*(3), 260–279.

Chan, W. W. L., Au, T. K., & Tang, J. (2014). Strategic counting: A novel assessment of place-value understanding. *Learning and Instruction, 29*(1), 78–94.

Collet, M. (2003). *Diagnostic assessment of the understanding of the base-ten-system.* Paper presented at the symposium current issues in assessment of learning disabilities of The Congress of the European Federation of Psychologists Associations (EFPA). Vienna.

Freudenthal, H. (1973). *Mathematics as an educational task.* Dordrecht, Holland: Reidel.

Freudenthal, H. (1991). *Revisiting mathematics education.* Dordrecht/Boston/London: Kluwer Academic Publishers.

Fung, C. I. (1999). Teaching for Mathematising: From the dream to the reality (in Chinese). In N.-Y. Wong & K.-M. Wong (Eds.), *Proceedings of the Symposium on the Optimisation of Basic Mathematics Education* (pp. 4–46). Hong Kong: Department of Curriculum and Instruction, Faculty of Education, Chinese University of Hong Kong.

Fung, C. I. (2004). Teaching for Mathematising: Theory, practice, and prospect (in Chinese). In K.-M. Tang, K.-L. Wong, M.-S. Lee & N.-C. Mok (Eds.), *Proceedings of Hong Kong Mathematics Education Conference 2004* (pp. 78–88). Hong Kong: Hong Kong Association for Mathematics Education.

Fuson, K. C., & Briars, D. J. (1990). Using a Base-Ten Blocks Learning/Teaching Approach for First- and Second-Grade Place-Value and Multidigit Addition and Subtraction. *Journal for Research in Mathematics Education, 21*(3), 180–206.

Geary, D. C. (2000). From infancy to adulthood: The development of numerical abilities. *European Child & Adolescent Psychiatry, 9*(Suppl. 2), II/11–II/16.

Han, Y., & Ginsburg, H. P. (2001). Chinese and English mathematics language: The relation between linguistic clarity and mathematics performance. *Mathematical Thinking and Learning, 3*(2&3), 201–220.

Haylock, D. (2011). Number and place value. In D. Haylock (Ed.), *Mathematics explained for primary teachers* (4th ed., pp. 65–82). The UK, London: Sage.

Haylock, D., & Cockburn, A. (2008). *Understanding mathematics for young children: A guide for foundation stage and lower primary teachers* (pp. 30–58). The UK, London: Sage.

Ho, C. S. H., & Cheng, F. S. F. (1997). Training in place-value concepts improves children's addition skills. *Contemporary Education Psychology, 22*(4), 495–506.

Lawton, F., & Hansen, A. (2011). Numbers and the number system. In A. Hansen (Ed.), *Children's error in mathematics: Understanding common misconceptions in primary schools* (2nd ed., pp. 20–46). The UK, Exeter: Learning Matters Ltd.

McGuire, P., & Kinzie, M. B. (2013). Analysis of place value instruction and development in pre-kindergarten mathematics. *Early Childhood Education Journal, 41*(5), 355–364.

Moeller, K., Pixner, S., Zuber, J., Kaufmann, L., & Nuerk, H. C. (2011). Early place-value understanding as a precursor for later arithmetic performance—A longitudinal study on numerical development. *Research in Developmental Disabilities, 32*(5), 1837–1851.

Munn, P. (2008). Children's beliefs about counting. In I. Thompson (Ed.), *Teaching and learning early number* (2nd ed., pp. 19–33). The UK, Buckingham: Open University Press.

Ng, S. S. N., & Rao, N. (2010). Chinese number words, culture, and mathematics learning. *Review of Educational Research, 80*(2), 180–206.

Ross, S. H. (1989). Parts, wholes, and place value: A developmental view. *The Arithmetic Teacher, 36*(6), 47–51.

Ryan, J., & Williams, J. (2007). Developing number. In J. Ryan & J. Williams (Eds.), *Children's mathematics 4–15: Learning from errors and misconceptions* (pp. 53–78). The UK, Maidenhead: Open University Press.

Sharma, M. C. (1993). Place value concept: How children learn it and how to teach it. *Math Notebook, 10*(1–2), 3–26.

Valeras, M., & Becker, J. (1997). Children's development understanding of place value: Semiotic aspects. *Cognition and Instruction, 15*(2), 265–286.

Van Oers, B. (2014). The roots of mathematising in young children's play. In U. Kortenkamp, B. Brandt, C. Benz, G. Krummheuer, S. Ladel, & R. Vogel (Eds.), *Early mathematics learning: Selected papers of the POEM 2012 conference* (pp. 111–123). The Netherlands, Dordrecht: Springer.

Authors Biography

Dr. Mun Yee Lai is a senior lecturer in mathematics at the Australian Catholic University. She has extensive experience and lectures in both primary and secondary mathematics education. She conducts research into pre-service teachers' mathematical knowledge for teaching, teaching for mathematizing, using Variation Theory in Learning Study and Chinese primary children's misconception in decimal numbers. Recently, she has begun to explore the application of Planned Behaviour Theory for understanding school learners' mathematics learning intention.

Dr. Chun-Ip Fung received his doctorate in 2001 at the University of Hong Kong and has been teacher educator for twenty-five years. He was the President of the Hong Kong Association for Mathematics Education from 2000 to 2004 and is currently an assistant professor of the Education University of Hong Kong. His research interest includes design science approach to mathematics education and mathematics teacher development.

Chapter 7
From Cradle to Classroom: Exploring Opportunities to Support the Development of Shape and Space Concepts in Very Young Children

Aisling Leavy, Jennifer Pope and Deirdre Breatnach

Abstract This chapter provides an overview of the growing body of evidence regarding the nascent geometrical capabilities of young children. We focus on spatial awareness and shape which are two broad concepts that underpin geometrical thinking from 0 to 8 years. In relation to each concept, we identify key indicators of informal learning trajectories, the mathematical underpinnings of these concepts and how they manifest themselves in early infancy through the preschool years and into the early primary school years. In parallel with these informal trajectories, we explain 'big ideas' which may include obstacles to learning or key misconceptions. Finally, we outline ways of supporting the development of geometrical understandings across the early years making reference to informal mathematics and play-based activities within the home and community setting and acknowledge the important role that language, gesture and context play in the development of children's geometrical capacities and understandings.

Keywords Geometry · Spatial awareness · Shape · Learning trajectories
Play · Gesture · Home · Community · Language · Birth to 8 years

A. Leavy (✉) · J. Pope · D. Breatnach
University of Limerick, Limerick, Ireland
e-mail: Aisling.Leavy@mic.ul.ie

J. Pope
e-mail: jennifer.pope@mic.ul.ie

D. Breatnach
e-mail: Deirdre.Breatnach@mic.ul.ie

V. Kinnear et al. (eds.), *Forging Connections in Early Mathematics Teaching and Learning*, Early Mathematics Learning and Development,
https://doi.org/10.1007/978-981-10-7153-9_7

7.1 Introduction

This chapter provides an overview of the growing body of evidence regarding the nascent geometrical capabilities of young children. Spatial awareness and shape are two broad concepts that underpin geometrical thinking from birth to 8 years. The interactions that children have from infancy and how they engage with their environment all serve to promote these geometrical understandings. Parents and educators working with very young children should be open and receptive to the opportunities that play a beneficial role in promoting broader mathematical concepts beyond a traditional focus on learning numbers or basic shape recognition. Adults should aim to 'create physical environments that are rich in exciting opportunities for mathematical activity' (Tucker, 2005, p. 15) and through active exploration 'maths will be a sense-making process with everyday relevance' (Diezmann & Yelland, 2000, p. 56).

In Ireland, *Aistear, The Early Childhood Curriculum Framework* (NCCA, 2009) highlights the continuum of learning from birth to the formal classroom for young children with one of the key themes being 'Exploring and thinking'. Aims within this key theme include to 'learn about and make sense of the world around them' with broad learning goals such as 'to develop a sense of time, shape, space and place' (NCCA, 2009, p. 44). In a similar vein, within *Belonging, Being & Becoming, the Early Years Learning Framework for Australia*, educators are encouraged to 'recognise mathematical understandings that children bring to learning and build on these in ways that are relevant to each child' (DEEWR, 2009, p. 35). How these understandings of shape and space concepts can be nurtured from birth and in the earliest years before school will be addressed throughout this chapter as well as exploring the opportunities for a deeper appreciation of the area of early geometry.

7.2 Spatial Awareness and Orientation (Location)

Spatial ability underpins the visual thinking required in all aspects of geometrical reasoning (Battista, 2007; NCTM, 2000). Furthermore, there is growing evidence of the links between spatial reasoning and other aspects of mathematics—specifically, early spatial awareness ability is related to early mathematics ability (van den Heuvel-Panhuizan, Elia, & Robitzsch, 2015) and acts as a predictor of early number knowledge at age 8 (Gunderson, Ramirez, Beilock, & Levine, 2012) and later mathematics achievement at grade 7 (Stannard, Wolfgang, Jones, & Phelps, 2001). The National Research Council [NRC] have noted that the trajectory of children's spatial development is reliant upon 'spatially relevant experiences, including those involving spatial language and spatial activities, such as block building, puzzle play, and experience with certain video games' (NRC, 2009, p. 72). Thus, fostering spatial awareness through appropriate experiences within the early years facilitates children's spatial development as well as broader mathematical competencies.

The spatial awareness of young children is primarily mediated through exploration of their immediate environment. This form of spatial awareness is termed 'spatial orientation', and it occurs as young children move physically from one area to another negotiating distance and direction as they move and play in indoor and outdoor spaces. Physically, infants will develop and reach milestones at an individual rate but within a general sequence (WHO, 2006), and as they develop, there are many activities that can support the development of spatial awareness. The following list presents opportunities for unrestricted movement which are important for developing spatial sense in very young infants:

- providing plenty of opportunities for floor play
- lying on their backs reaching for their feet (trying to suck their own toes!)
- trying to grasp interesting objects hanging overhead
- during tummy time, encouraging them to lift their heads by placing interesting objects slightly raised so that they need to make an effort to see them (but the level of challenge is appropriate) or a low mirror to see themselves peering in
- experimenting trying to roll over and back again
- pulling themselves to sit
- ball pools to sit in and later jump in and then to crawling or bottom-shuffling
- crawling or bottom-shuffling can be encouraged by having something to move towards or an interesting pathway to follow different textures, for example as simple as bubble wrap taped to the ground, a tunnel to crawl through (even an empty opened cardboard box) or messy experiences on the floor with gloop (cornflour and water mix) or edible paint on large taped sheets of paper for babies to explore on all fours or sitting
- having a box of different shaped and textured balls, for example that might roll in different ways or distances, has significant learning potential not only in terms of experimentation but also to chase after. Similarly, filling empty water bottles with different ingredients such as rice or varying volumes of water with the addition of glitter, drops of food colouring, etc., rolling the bottles will have different effects and retrieving the bottles will encourage not only physical development but also spatial awareness
- providing soft blocks to climb over as they become more adept crawlers
- providing bars by mirrors to pull themselves up to stand
- providing a baby trolley with wooden blocks to push along (baby walkers with castors that babies sit in have significant safety implications and are not recommended (Murphy & Nicholson, 2011)
- push and pull along toys or even emptied old handbags to drag around, doll prams, pushchairs or miniature shopping trollies
- older toddlers enjoy vehicles to sit on and trikes particularly if they have a trailer attached so that they can transport objects around with them a common schema (Athey, 2007)

It has been highlighted by the NRC that within spatial awareness, 'developmental rates and the competencies achieved are highly dependent on access to spatial activities, spatial language, and learning opportunities at home and at school' (NRC, 2009, p. 78). In line with other aspects of early mathematical learning, spatial concepts are strengthened through the provision of a rich environment that incorporates a variety of hands-on resources and manipulatives. These rich environments are critical as young children cannot go 'straight to abstractions, they need to handle and manipulate objects' (Knaus & Featherstone, 2014, p. 12) and build up a network of meaningful connections (Tucker, 2005). Spatial exploration also entails the use of finer motor skills such as those involved in the manipulation of puzzle pieces and geometric shapes into various positions within objects and containers. In infancy, this begins with manipulation of the first objects to hand—often their own hands which are a source of fascination. The educationalist Froebel (1782–1952), founder of the kindergarten, identified specific 'gifts' that young children should be presented in sequence from concrete objects to pictures to symbolic representations to facilitate their ability to make connections about the world with a strong emphasis on mathematical thinking. The first gift for an infant was a soft woollen ball (sometimes now sold in a series of woollen balls) in a solid colour that was the right size for a baby to hold, to drop, to mouth, to roll and to squash (and subsequently return to form). It was also designed with string so that the adult could move it. The second gift was a wooden sphere, a cube and a cylinder. Later gifts then include 8 cubes, longer building blocks and more cubes divided into halves and quarters (Tovey, 2013). Famously, the renowned architect Frank Lloyd Wright credited his success to the early introduction of these gifts to him as a small child.

To address the potentially more clinical environment of a day care setting and to bridge the gap between a home environment, the treasure basket as pioneered by Goldschmeid and Jackson (1994) is a very useful resource for infants within both the early childhood setting and within the home environment. Designed for infants when they are able to sit up but before they are mobile, the treasure basket aims to provide an opportunity for self-directed exploration for the infant. According to Goldschmeid's original specifications, a treasure basket should be over 14 in. (35 cm) in diameter, four to five inches high, flat bottomed with no handles and strong enough for an infant to lean on. The treasure basket should be filled to the brim with natural everyday household items with plastic being avoided. Each item selected should be safe for the infant to mouth, and the items included should aim to stimulate a variety of senses and could include the following: a lemon, golf ball, large pine cone, stainless steel egg cup, wooden spoon, loofah, large wooden curtain rings, etc. The infant (or infants) explores the contents uninterrupted with the adult close by as a physical support but not disturbing the infant in deep concentration taking items out, putting items in, etc. The aim is that young infants will consider the different properties of the items through a variety of senses, often returning to the same items or repeating the same procedures such as banging the curtain ring off the side of the basket. This rudimentary experience allows for rich learning, deep concentration and an opportunity to explore space and shape in a very hands-on way. The infant is in control and the experience ends when the infant signifies the end, the adult initiates the activity but does not lead it, and

Goldschmeid and Jackson (1994) argued that by interferring, the adult can be more detrimental to the learning cycle.

For the toddler, this type of heuristic (exploratory) play with particular objects can be developed to include several of the same object, obviously useful for counting, matching and sorting but for understanding space and shape also. In Fig. 7.1, you can see several bun cases arranged or an empty muffin tin with 12 cups to fill, a collection of washing pegs or arranging several books as a tunnel (Fig. 7.2). Nutbrown (2011, p. 86) argues that 'mathematical ideas are not only present in children's choices and selections of materials but also in the ways they use the materials available to them', and it is the use of the materials that 'can impact on the learning potential of what is done'. She also argues that children explore space in a flexible way and questions the idea of only providing children with rigid, limited opportunities to explore space such as jigsaws and asks that we consider how we provide activities that support the concept of flexible space. Manipulating gloop, sand, play dough or baking and preparing food are other useful experiences.

Research has also suggested that a child's emerging spatial concepts are enhanced by the use of rich language appropriate to the specific spatial experience (NRC, 2009; NCCA, 2014; Pound, 2006). With regard to spatial language, it has been noted that in laboratory experiments where children were given specific

Fig. 7.1 Spatial arrangement of bun cases [Author's own image]

Fig. 7.2 Arranging books as a tunnel [Author's own image]

linguistic spatial words when hiding certain objects such as 'I'm putting this on/in/under the box', these children were better able to find the objects than children who heard a general reference to location—'I'm putting this here' (Gentner, 2003 cited in NRC, 2009, p. 76). In studies where parental use of spatial language was encouraged and their children were subsequently involved in spatial tasks, it was concluded that 'children's spatial language production is a significant predictor of their later spatial skills' (Pruden, Levine, & Huttenlocher, 2011). Copley (2010, p. 114) has recommended that frequent usage of the following terms within authentic experiences supports young children's understanding of spatial concepts:

Spatial Vocabulary

Location/position words: on, off, on top of, over, under, in, out, into, out of, top, bottom, above, below, in front of, in back of, behind, beside, by, next to, between, same/different side, upside down

Movement words: up, down, forward, backward, around, through, to, from, toward, away from, sideways, across, back and forth, straight/curved path

Distance words: near, far, close to, far from, shortest/longest path

Transformation words: turn, flip, slide

Once children pass through infancy and become more independent and mobile, the opportunities for developing spatial sense increase. The following are opportunities for play-based activities in the preschool years which target spatial awareness and orientation: outdoor activities which involve large-scale apparatus that toddlers and preschool children can exploit in various ways; activities that support developing spatial sense include climbing up ladders and over obstacles (Fig. 7.3); crawling through tunnels (Fig. 7.4); manoeuvring dolls' buggies and other wheeled vehicles around and between other objects.

It is of key importance that early years' educators use the rich spatial vocabulary outlined above in an informal and playful way during these activities. This kind of exploration supports the young child's 'spatial orientation' as ideas such as direction (which way?), distance (how far?), location (where?), and identification (what objects?) are highlighted.

- Open-ended materials such as large cardboard boxes (Fig. 7.5), hollow blocks and planks allow children to 'build structures big enough to get inside, which allows them to experience their constructions from a very different spatial perspective' (Cartwright, 1996 cited in Copley, 2010, p. 115).
- Small world play—resources such as farm or zoo play sets allow children to move characters around various scenes and allow them to develop narratives in which positional language can be utilised (Tucker, 2010). Spatial visualisation through perspective taking can also be explored within small world play by encouraging a child to describe what he/she sees from different viewpoints when engaging with other children in small world play
- Certain stories, such as 'The Three Billy Goats Gruff', can introduce spatial vocabulary in an amusing and appealing way (Fig. 7.6), and provide compelling contexts for young learners (Hourigan & Leavy, 2016).

Fig. 7.3 Navigating obstacles. Image source: creativecommons.org available at https://commons.wikimedia.org/wiki/File:Outdoor_Play_area.jpg

Fig. 7.4 Crawling through a tunnel [Author's own image]

- Initiating spatial visualisation activities by utilising various forms of puzzle play such as pictorial jigsaws or shape-based insets similar to those within a Montessori-based approach (Fig. 7.7) supports spatial awareness.
- Transforming 3-D and 2-D shapes by engaging in games such as 'Sorting Parcels' with 3-D shapes (Tucker, 2010) provide valuable experiences. A similar activity involves transforming 2-D shapes through folding or cutting and reconstructing to form new shapes. Figure 7.8 is an example of dissecting squares to produce triangles and parallelograms.

The insights provided by research into the sophistication of preschool children's spatial reasoning provide support for recent recommendations that school-based geometry instruction focuses initially on spatial geometry rather than the traditional emphasis on plane geometry (Clements & Sarama, 2004; Goldenberg, Clements, Zbiek, & Dougherty, 2014). Studies of spatial reasoning in this age group focus on perspective taking and mental rotation (van den Heuvel-Panhuizan, Elia, & Robitzsch, 2015) and indicate that that most children achieve competence in perspective taking somewhere between age 5–8 (Flavell, Abrahams Everett, Croft, &

Fig. 7.5 Exploring a three-dimensional structure from inside. Image source: freedigitalphotos.net available at http://www.freedigitalphotos.net/ Miles, S, 2011—image ID: 10053849

Flavell, 1981; Pillow & Flavell, 1986). Evidence of the presence of mental rotation abilities exists in infants as young as 22-month-olds (Örnkloo & von Hofsten, 2007) and particularly among 4–8-year-olds (Bruce & Hawes, 2015; Casey et al., 2008). These latter studies suggest that, given the correct conditions, children's mental rotation skills can be improved upon in the early school years.

Acknowledgement of the fundamental role played by gestures in mathematical communication and in mediating meaning extends to spatial thinking. Gestures present an important means of communicating abstract ideas (Cook & Goldin-Meadow, 2006), prior to the use of language and formal mathematics, for young children. Examination of the gestures of young learners during activities that entail communication about space indicates that 5-year-olds use gestures to communicate meanings. Multidimensional relationships between speech and gestures exist (Elia, Gagatsis, Michael, Georgiou, & van den Heuvel, 2014); at times, the meanings of 5-year-olds gestures and speech match, and at other times, gestures replace speech, and at other times, gestures complement the meanings expressed in speech. Thus, gestures have the potential to improve early spatial skills and contribute to the construction of mathematics knowledge (Roth & Lee, 2004). Furthermore, young learners are aware of the gestures of others, draw information from them (Ping & Goldin-Meadow, 2008) and possibly enhance their geometric

Fig. 7.6 Using story to develop spatial awareness and orientation

Fig. 7.7 Shape-based inserts similar to those advocated by Montessori practitioners. Image source http://www.lisheenmontessori.com/product_view.php?id=266

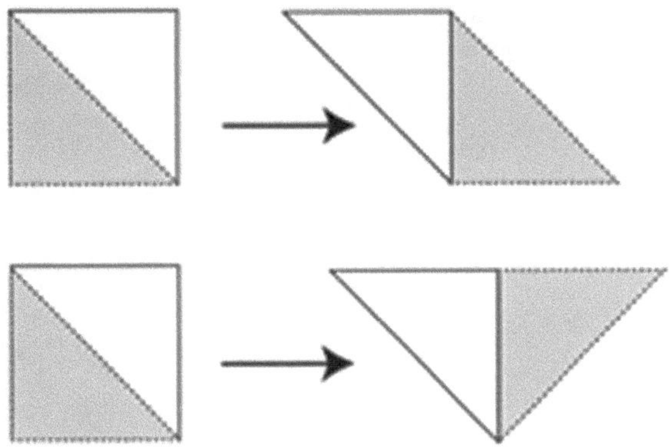

Fig. 7.8 Dissecting and transforming squares to produce new shapes

learning as a result of observing the gestures of others (Elia et al., 2014; Valenzeno, Alibali, & Klatzky, 2003).

School-based activities in the early years build on preschool activities and expand focus by engaging children in the following recommended activities:

- Experiences *outside the classroom walls* are advocated for the development of spatial orientation concepts, in particular the use of maps and plans (Thom & Garvey, 2015) and the use of *virtual tours* (Hourigan & Leavy, 2015). Similarly, the use of perspective-taking activities can also maximise the potential of contexts outside the classroom.
- The use of *construction activities* and experiences with models of geometric objects supports the development of spatial reasoning and mental rotation (Bruce & Hawes, 2015; Thom & Garvey, 2015).
- The use of *structured maps* (DeLoache, Pierroutsakos, Uttal, Rosengren, & Gottlieb, 1998) and narratives of *shrinking machines* (DeLoache, Miller, & Rosengren, 1997), which involve reducing a room to the size of a given model, has been shown to bring about improvements in spatial skills which are utilised in navigation and map-reading activities.
- The development of spatial awareness is also enhanced through the use of *technology environments* combined with physical movement and pen-and-pencil activities (Sarama & Clements, 2009)

7.3 Shape

There are multiple opportunities for manipulating shapes in the infant years. These opportunities go beyond shape sorters and cutters and can involve heuristic play opportunities as described earlier (such as the ball box, Figs. 7.9, 7.10 and 7.11). The usual toys such as shape sorters, early shape jigsaws and shape stackers still play a role but if the play equipment is expanded to include everyday utensils

Fig. 7.9 Child exploring the ball box [Author's own image]

Fig. 7.10 Child exploring the ball box [Author's own image]

(Figs. 7.12 and 7.13), it then contextualises the experience for infants and toddlers and has the potential to facilitate the ability to draw inferences later on because the range of their experiences and contexts are increased.

Fig. 7.11 Playing inside the ball box [Author's own image]

Fig. 7.12 Exploring an everyday object to develop a sense of shape [Author's own image]

The knowledge and understanding of shape which children display in the pre-school years reflects predominantly the first level of Van Hiele's geometric thinking development, i.e. visualisation (Van Hiele, 1986). As suggested by the title,

Fig. 7.13 Exploring everyday objects to develop a sense of shape [Author's own image]

children at the 'visualisation' level view shapes by their appearance. Thus, the perception of an object informs the child's thinking. Or, as stated by the NRC, 'a given figure is a rectangle, for example, because "it looks like a door"' (NRC, 2009, p. 176). Children at this level do not think about the particular characteristics of shapes rather they see them as wholes, and they rely on perceptual input to support their understanding; such limited understanding can be prevalent in children as old as 10 (Hourigan & Leavy, 2015). As progression through each of Van Hiele's levels depends on understanding of a previous level (Hock, Rohani, Aida, Yunus, & Ahmad, 2015), it is vital that children's geometric experiences in their preschool years provide them with the foundation for future learning. Furthermore, as noted by Copley (2010), educators also have a key role in connecting children's playful experiences to fundamental geometric concepts and other mathematical ideas.

> Familiarity with shape, structure, location, and transformations and development of spatial reasoning enable children to understand not only their spatial world but also other mathematics topics. As children count the sides of two-dimensional shapes or the faces of a cube, they learn about number relationships. Patterns, functions, and even rudiments of algebra may be noted when children identify patterns in space or when they see the relationships between the number of faces, edges, and vertices of three-dimensional figures. When children compare shapes, directions, and positions in space, they develop concepts and acquire vocabulary that they also put to use in measurement. Grouping items, sometimes by shape or another geometric feature, is a skill also fundamental to data collection, and children may record and report shapes in an activity or in the environment. (p. 106)

A key idea within geometry is that shapes have different parts and aspects that can be described, and they can be composed and decomposed. While children at preschool age rely heavily on perception and concrete experiences to support their knowledge of 2-D and 3-D shapes, they do begin to think about the components of shapes and how the composition relates to the entire figure. The NRC (2009) has noted that children's understanding of 2-D- and 3-D-shape concepts emerges over time. It has outlined phases of sophistication in children's thinking at three levels: thinking visually/holistically, thinking about parts, and relating parts and wholes.

For example, children up to 2-year-olds might manifest the beginnings of thinking visually/holistically through the rudimentary stacking of 3-D shapes. At 3 years of age, thinking about parts could be evidenced by the child differentiating between 2-D and 3-D shapes intuitively, marked by accurate matching or naming. Then, at 4 years of age, relating parts to wholes could be displayed through the child's description of why some blocks 'stack well' and others do not or through the systematic creation of building blocks to produce composite shapes such as arches, enclosures, corners and crosses.

In the design of curricular content to support geometric concept development within the preschool context, due consideration should be given to providing playful and engaging activities which stimulate young children's curiosity. Indeed, Clements and Sarama (2009) contend as children represent and reflect on play-based experiences that they become mathematical. The early year's educator has a key role in both 'pedagogical framing' and 'pedagogical interactions' to support the child's learning of geometry (Siraj-Blatchford, 2004). This 'framing' of the learning environment by the early year's educator ensures that children have varied experiences of geometric concepts. Hewitt (2001) notes that young children gain an understanding of the properties of 3-D shapes through a process of taking apart, examining and rebuilding their constructions. While with appropriate 'pedagogical interactions', children can be encouraged for example to discuss their 3-D creations, thus allowing the educator to support the bridging of children's informal descriptions with the formal language of shape, 'the teacher might decide to provide the word sphere for the basketball and engage Jeffrey in considering other differences between a ball and a two-dimensional circle' (Copley, 2010, p. 106). Interactions such as these support movement through the different phases outlined in Table 7.1.

In addition, when extensive experiences of manipulation of 3-D shapes are provided for preschool children, these explorations assist in their understanding of properties of 2-D formats. As Pound (2006) notes the physical form of 3-D shapes assists children in describing 2-D shapes as the faces and sides are more easily identifiable from the 3-D format. Pedagogical interactions which use the specific language to highlight the geometric experience also facilitate the different levels of thinking outlined in Table 7.1. Additional considerations and activities to support the development of shape in the early years include:

- The provision of a variety of large-scale and small-scale blocks in early years' environment to support children's exploration of 3-D materials (Fig. 7.14)
- Incorporating large periods of time into the play schedule to allow children to create varying constructions
- Extending children's thinking by using descriptions, comments and questions on children's creations, e.g. 'this reminds me of a crane' or 'tell me how you made this crane'
- The use of accurate terminology and bridging the informal language of children with formal mathematical language allows children to maximise the mathematical potential through play with manipulatives.

Table 7.1 Connections between spatial experiences, thinking and development. Adapted from NRC (2009)

Approximate age	2-D shapes	3-D shapes	Spatial awareness	Level of thinking
0–2 years	Starting to form unconscious visual schemes for shapes	Shape sorters-matching 2-D and 3-D shapes Rudimentary stacking	Using distance and direction when they move Learning relational language (in, out, etc.) and vertical discretion—up and down	Beginning thinking visually/holistically
At 2–3 years	Recognises and informally describes initially circle and square	Name a side of a cube a square Basic building with similar shapes to represent an object	Remembering location and using landmarks and distance between them	Thinking visually/holistically
At 3 years	Names and describes 2-D shapes particularly, circle and square Triangle and rectangles recognised in 'prototypical' formats	Used extensively in block play—but tend to be named according to their 2-D faces Block play may include vertical and horizontal structures as well as constructions in multiple directions, possibly creating arches, enclosures, corners and crosses	Uses positional language such as 'in', 'out', 'under' and along with such vertical directionality terms as 'up' and 'down'	Thinking visually/ holistically
At 3 years		Represent real-world objects with blocks that have a similar shape Combine unit blocks by stacking		Thinking visually/ holistically
At 3 years	Differentiates between 2-D and 3-D shapes intuitively, marked by accurate matching or naming.			Thinking about parts
At 3 years	Begins to describe shapes by number of sides			Thinking about parts
At 4 years	Recognises and describes informally multiple orientations, sizes and shapes (includes circles and half/quarter circles, squares and rectangles, triangles and others [the pattern block rhombus, trapezoids, hexagons regular])		Uses relational words of proximity, such as 'beside', 'next to', 'between', 'above', 'below', 'over' and 'under'	Thinking visually/ holistically

(continued)

Table 7.1 (continued)

Approximate age	2-D shapes	3-D shapes	Spatial awareness	Level of thinking
At 4 years	Describe the difference between 2-D and 3-D shapes, and names common 3-D shapes informally and with mathematical names ("ball"/sphere; 'box' or rectangular prism, 'rectangular block', or 'triangular block'; 'can'/cylinder)		Uses relational words of proximity, such as 'beside', 'next to', 'between', 'above', 'below', 'over', and 'under'	Thinking visually/ holistically
At 4 years	Identify faces of 3-D objects as 2-D shapes and name those shapes		Use relational language involving frames of reference such as 'in front of', 'back of', 'behind' and 'before'	Thinking about parts
At 4 years	Identify (matches) the faces of 3-D shapes to (congruent) 2-D shapes, and match faces of congruent 2-D shapes, naming the 2-D shapes		Combine building blocks, using multiple spatial relations	Thinking about parts
At 4 years	Represent 2-D and 3-D relationships with objects			Thinking about parts
At 4 years		Informally describe why some blocks 'stack well' and others do not		Relating parts and wholes
At 4 years		Compose building blocks to produce composite shapes. Produce arches, enclosures, corners and crosses systematically		Relating parts and wholes

Fig. 7.14 3-D blocks and shapes to build structures. Image source: creativecommons.org https://www.flickr.com/photos/66992990@N00/with/8381414364/

Prior to school age, when engaging in concept formation, children may be able to recognise shape names without being aware of the characteristics or properties that make up that concept. Furthermore, they may not be able to identify or construct non-examples of the construct. This is known as partitive classification (Clements & Sarama, 2007) and equates with the Van Hiele Level 0 classification known as visualisation (see Table 7.1).

Children in the early school years develop mental constructs of mathematical concepts through abstraction from examples (Skemp, 1969). This has ramifications for the types of examples used to represent particular shapes. A rich set of examples which distinguish between the essential and non-essential characteristics of a shape is necessary. Studies with young children have revealed the power of prototypes (common or typical exemplars) in influencing the acceptance or rejection of a specific geometric figure as belonging to a class of shapes (Hershkowitz, 1989). Moreover, there is evidence to suggest that over use of prototypical examples may lead children to focus on irrelevant aspects of examples, may lead to difficulties in recognising particular shapes as belonging to a specific category and may result in making assumptions about shapes that are not correct (Fox, 2000; Hourigan & Leavy, 2015). For example, the first triangle in Fig. 7.15, the equilateral triangle, is a common triangle prototype. However, if children are exposed only to this prototype, they may not recognise that the other shapes in Fig. 7.15 are also triangles.

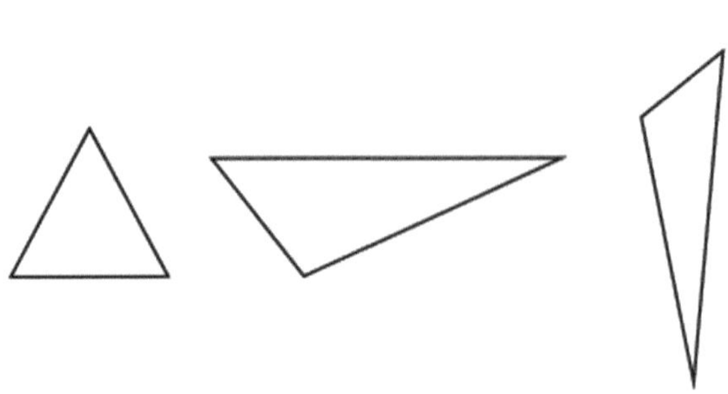

Fig. 7.15 Prototypical and non-prototypical triangles

A salient example is when we presented a three-year old with the third triangle in Fig. 7.15 and asked her to identify it—she replied that it was 'a paper aeroplane'!

Emphasis has been placed on the use of non-examples as they provide information about what is and what is not included in the definition of a concept (Wilson, 1986). Thus, non-examples focus children on the essential characteristics of the concept and delineate the boundaries of the concept (Charles, 1980). The shape in Fig. 7.16 is a useful non-example as it contains many of the essential characteristics of a triangle (3 sides, 3 angles). At the same time, this non-example delineates the boundaries between polygons and non-polygons. A triangle is a polygon and hence has straight sides, a feature that the non-example is Fig. 7.16 does not possess. Non-examples can also be fruitfully use in shape-sorting activities and the use of property loops (see Hourigan & Leavy, 2015).

As is the case with spatial reasoning, research in shape reveals the importance of not limiting attention solely to the use of language and recommends attention to the use of gestures. The importance of gesture remains even when formal mathematical language is developed as evidenced in a study of 7-year-olds by Kaur (2015) which found that speech only was not sufficient for children when communicating about shape in a dynamic geometric environment. Children in this study used head-turning gestures, arm-stretching gestures and finger gestures to communicate visual and dynamic information about triangles. In addition to the constructive use of gestures and experiences with a range of non-prototypical shapes, the following activities support understandings of shape in the early years of formal schooling:

- In an examination of experiences provided to children in relation to 2-D and 3-D shapes, Kaur (2015) cautions that young children have few experiences with 1-D objects and proposes the provision of experiences drawing lines with a view to highlighting the relationship with sides and edges as they pertain to 2-D and 3-D shapes.
- The role of *active participation* is emphasised in recommendations to support the development of geometrical reasoning. The construction of geometric forms using craft materials (such as construction sticks, matchsticks, straws, card,

Fig. 7.16 Non-example of a triangle

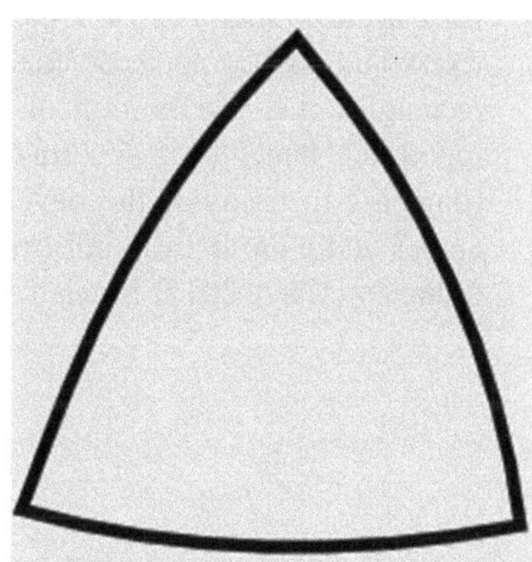

glue) and more structured materials (such as geoboards) supports the development of young learners' concept definitions and to precipitate a move from the use of everyday language to the use of explicit geometric terminology (Leavy & Hourigan, 2015; Pytlak, 2015).

- The focus on active construction has led to increasing awareness of the *limitations of textbooks* in geometry instruction (Hallowell, Okamoto, Romo, & La Joy, 2015) and the concomitant cognitive demands made on young children to extract spatial relations within and across shapes. Interrogation of use of diagrams presented within textbooks provides evidence to suggest that *how* a geometric shape is constructed and by *whom* (Thom & McGarvey, 2015) influences the ability of children to use the diagram. In other words, the benefit of children constructing their own shapes far outweighs the passive experience of working with textbooks.
- Use of *shape-sorting activities* provides insights into whether pupils identify essential and non-essential characteristics of shapes (Hourigan & Leavy, 2015). Young children may choose shapes based on topological properties, such as orientation and appearance, and neglect to take account of the properties of the shape. For example, if presented with the triangles in Fig. 7.15, many children may not sort the second and third triangles into the group of triangles due to their orientation, i.e. as they do not sit on a horizontal base. The importance of *group discussion* during shape-sorting activities supports pupils in reflecting on categorisations and essential properties and allows them to clarify concepts and supports the development of strong concept images (Rodrigues & Serrazina, 2015).
- Increasing attention has been placed on the *role of technology*, and dynamic geometry environments, in supporting the development of young learners' geometrical understandings. The use of static media, text and textbooks, alone to represent geometrical concepts may limit what children can produce as compared to more dynamic media (Kaur 2015). The use of programming robots to move along paths has been beneficial in a study of 7–8-year-olds in terms of seeing paths as representing the boundaries of figures and in the subsequent exploration of the geometric characteristics of those figures (Bussi & Baccaglini-Frank, 2015). This study found that experience programming the robot laid the cornerstone for the development of an inclusive definition of rectangles (i.e. a definition of rectangles that included squares). Two studies report the benefits of situating the exploration of triangles in dynamic environments in terms of the development of understandings of properties of triangles and a move from informal descriptions to the use of formal properties of triangles (Kaur, 2015; Sinclair & Moss, 2012).

7.4 Conclusion

This chapter was entitled 'From Cradle to Classroom' in order to outline some key indicators of informal learning trajectories within the concepts of shape and space and to identify the mathematical underpinnings of these concepts and how they manifest themselves from early infancy through the preschool years and into the early primary school years. In parallel with these informal trajectories, 'big ideas' were discussed including obstacles to learning or key misconceptions. Finally, supporting the development of geometrical understandings across the early years with reference to informal mathematics and play-based activities within the home and community setting was acknowledged as well as the important role that language, gesture and context play in the development of children's geometrical capacities and understandings. Recognising the continuum of learning and the capacity that young children have to learn and knowing how mathematical concepts develop from infancy is the first step. Seeing the learning potential of everyday opportunities, the 'breadth and depth of experience' (Nutbrown, 2011, p. 87) and then extending learning within such contexts that are relevant to children is the challenge but also the marker of an effective early childhood educator.

References

Athey, C. (2007). *Extending thought in young children*. London: Sage.

Battista, M. T. (2007). The development of geometric and spatial thinking. In F. Lester (Ed.), *Second handbook of research on mathematics teaching and learning* (pp. 843–908). Reston: National Council of Teachers of Mathematics.

Bruce, C., & Hawes, Z. (2015). The role of 2D and 3D mental rotations in mathematics for young children: What is it? Why does it matter? And what can we do about it? *ZDM Mathematics Education, 47*(3), 331–343.

Bussi, M. G., & Baccaglini-Frank, A. (2015). Geometry in early years: Sowing seeds for a mathematical definition of squares and rectangles. *ZDM Mathematics Education, 47*, 391–405.

Cartwright, S. (1996). Learning with large blocks. In E. S. Hirsch (Ed.), *The block book* (3rd ed.). Washington, DC: National Association for Education of Young Children (NAEYC).

Casey, B. M., Andrews, N., Schindler, H., Kersh, J. E., Samper, A., & Copley, J. (2008). The development of spatial skills through interventions involving block building activities. *Cognition and Instruction, 26*(3), 269–309.

Charles, R. (1980). Exemplification and characterization moves in the classroom teaching of geometry concepts. *Journal for Research in Mathematics Education, 11*(1), 10–21.

Clements, D., & Sarama, J. (2004). *Engaging young children in mathematics: Standards for early childhood mathematics education*. Mahwah: Erlbaum.

Clements, D., & Sarama, J. (2007). The development of geometric and spatial thinking, students and learning. In F. K. Lester (Ed.), *Second handbook of research on mathematics teaching learning* (pp. 489–517). Reston, VA: NCTM.

Clements, D., & Sarama, J. (2009). *Learning and teaching early math: The learning trajectories approach*. New York, NY: Routledge.

Cook, S., & Goldin-Meadow, S. (2006). The role of gesture in learning: Do students use their hands to change their minds? *Journal of Cognition and Development, 7*(2), 211–232.

Copley, J. (2010). *The young child and mathematics* (2nd ed.) Washington: National Association for the Education of Young Children (NAEYC).

DeLoache, J. S., Miller, K. F., & Rosengren, K. S. (1997). The credible shrinking room: Very young children's performance with symbolic and non-symbolic relations. *Psychological Science, 8*, 308–313.

DeLoache, J. S., Pierroutsakos, S. L., Uttal, D. H., Rosengren, K. S., & Gottlieb, A. (1998). Grasping the nature of pictures. *Psychological Science, 9*, 205–210.

Department of Education, Employment and Workplace Relations [DEEWR]. (2009). *Belonging, being & becoming—The early years learning framework for Australia.* Canberra, ACT: Commonwealth of Australia.

Diezmann, C., & Yelland, N. J. (2000). Developing mathematical literacy in the early childhood years. In N. J. Yelland (Ed.), *Promoting meaningful learning* (pp. 47–58). Washington: NAEYC.

Elia, I., Gagatsis, A., Michael, P., Georgiou, A. S., & van den Heuvel, M. (2014). The role of gestures in making connections between space and shape aspects and their verbal representations in the early years: Findings from a case study. *Mathematics Education Research Journal, 26*, 735–761. doi:10.1007/s13394-013-0104-5.

Flavell, J. H., Abrahams Everett, B., Croft, K., & Flavell, E. R. (1981). Young children's knowledge about visual perception: Further evidence for the level 1–level 2 distinction. *Developmental Psychology, 17*(1), 99–103.

Fox, T. B. (2000). Implications of research on children's understanding of geometry. *Teaching Children Mathematics, 6*, 572–576.

Gentner, D. (2003). Why we're so smart. In D. Gentner & S. Goldin-Madow (Eds.), *Language in mind: Advances in the study of language and thought* (pp. 195–235). Cambridge: MA: MIT Press.

Goldenberg, E. P., Clements, D., Zbiek, R. M., & Dougherty, B. (2014). *Developing essential understanding of geometry and measurement for teaching mathematics in pre-K–grade 2.* Reston, VA: NCTM.

Goldschmeid, E., & Jackson, S. (1994). *People under three: Young children in daycare* (1st ed.). London: Routledge.

Gunderson, E. A., Ramirez, G., Beilock, S. L., & Levine, S. C. (2012). The relation between spatial skill and early number knowledge: The role of the linear number line. *Developmental Psychology, 48*(5), 1229.

Hallowell, D. A., Okamoto, Y., Romo, L. F., & La Joy, J. R. (2015). First-graders' spatial-mathematical reasoning about plane and solid shapes and their representations. *ZDM Mathematics Education, 47*, 363–375.

Hershkowitz, R. (1989). Visualization in geometry: Two sides of the coin. *Focus on Learning Problems in Mathematics, 11*(1), 61–76.

Hewitt, K. (2001) Blocks as a tool for learning: *Historical and Contemporary Perspectives in Young Children 56* (1): 6–11. Washington: National Association for the Education of Young Children (NAEYC).

Hock, T. T., Rohani A. T., Aida S. M., Yunus, & Ahmad, F. (2015). Understanding the primary school students' van Hiele levels of geometry thinking in learning shapes and spaces: A Q-methodology. *Eurasia Journal of Mathematics, Science & Technology Education, 11*(4), 793–802.

Hourigan, M., & Leavy, A. (2015). What's a real shape? Designing appropriate geometric instruction. *Australian Primary Mathematics Classroom, 20*(1), 24–29.

Hourigan, M., & Leavy, A. M. (2016). Practical problems: Introducing statistics to Kindergarteners. *Teaching Children Mathematics, 22*(5), 283–291.

Kaur, H. (2015). Two aspects of young children's thinking about different types f dynamic triangles: Prototypicality and inclusion. *ZDM Mathematics Education, 47*, 407–420.

Knaus, M., & Featherstone, S. (2014). *Maths is all around you.* London: Featherstone.

Leavy, A., & Hourigan, M. (2015). Budding architects: Exploring the designs of pyramids and prisms. *Australian Primary Mathematics Classroom, 20*(3), 17–23.

Murphy, A., & Nicholson, A. J. (2011). Baby walkers in Europe-time to consider a ban. *Irish Medical Journal, 104*(3), 69.

National Council for Curriculum and Assessment (NCCA). (2009). *Aistear: The early childhood curriculum framework. Principles and themes*. Dublin, Ireland: Author.

National Council for Curriculum and Assessment (NCCA). (2014). *Mathematics in early childhood and primary education*. Dublin, Ireland: Author.

National Council of Teachers of Mathematics. (2000). *Principles and standards for school mathematics*. Reston: NCTM.

National Research Council (NRC). (2009). Mathematics learning in early childhood: Paths towards excellence and equity. In C. Cross, T. Woods, & H. Schweingruber (Eds.), *Committee on Early Childhood Mathematics, Center for Education, Division of Behavioral and Social Sciences*. Washington, DC: The National Academies Press.

Nutbrown, C. (2011). *Threads of thinking*. London: Sage.

Örnkloo, H., & von Hofsten, C. (2007). Fitting objects into holes: On the development of spatial cognition skills. *Developmental Psychology, 43*(2), 404–416.

Pillow, B., & Flavell, J. (1986). Young children's knowledge about visual perception: Projective size and shape. *Child Development, 57*, 125–135.

Ping, R., & Goldin-Meadow, S. (2008). Hands in the air: Using ungrounded iconic gestures to teach children conservation of quantity. *Developmental Psychology, 44*(5), 1277–1287.

Pound, L. (2006). *Supporting mathematical development in the early years*. Oxford: Open University Press.

Pruden, S. M., Levine, S. C., & Huttenlocher, J. (2011). Children's spatial thinking: Does talk about the spatial world matter? *Developmental Science, 14*(6), 1417–1430. doi:10.1111/j.1467-7687.2011.01088.x.

Pytlak, M. (2015). Learning geometry through paper-based experiences. *Conference proceedings of Congress of European Research in Mathematics Education (CERME9)*. Prague.

Rodrigues, M.P., & Serrazina, L. (2015). Six years old pupils intuitive knowledge about triangles. *Conference proceedings of Congress of European Research in Mathematics Education (CERME9)*. Prague.

Roth, W. M., & Lee, Y. J. (2004). Interpreting unfamiliar graphs: A generative, activity-theoretic model. *Educational Studies in Mathematics, 57*, 265–290.

Sarama, J., & Clements, D. H. (2009). *Early childhood mathematics education research. Learning trajectories for young children*. New York: Routledge.

Sinclair, N., & Moss, J. (2012). The more it changes, the more it becomes the same: The development of the routine of shape identification in dynamic geometry environments. *International Journal of Education Research, 51&52*, 28–44.

Siraj-Blatchford, I. (2004). Quality teaching in the early years. In A. Anning, J. Cullen, & M. Fleer (Eds.), *Early childhood education: Society and culture* (pp. 137–148). London, Sage Publications.

Skemp, R. (1969). *The psychology of learning mathematics*. Harmondsworth, UK: Penguin.

Stannard, L., Wolfgang, C. H., Jones, I., & Phelps, P. (2001). A longitudinal study of the predictive relations among construction play and mathematical achievement. *Early Child Development and Care, 167*(1), 115–125.

Thom, J.S., & McGarvey, L. (2015). The act and artifact of drawing(s): Observing geometric thinking with, in, and through children's drawings. *ZDM Mathematics Education (Special issue: Geometry in primary school), 47*(3), 465–481.

Tovey, H. (2013). *Bringing the Froebel approach to your early years practice*. New York: Routledge.

Tucker, K. (2005). *Mathematics through play in the early years*. London: Paul Chapman.

Tucker, K. (2010). *Mathematics through play in the early years* (2nd ed.). London: Paul Chapman.

Valenzeno, L., Alibali, M. A., & Klatzky, R. (2003). Teachers' gestures facilitate students' learning: A lesson in symmetry. *Contemporary Educational Psychology, 28*, 187–204.

Van den Heuvel-Panhuizen, M., Elia, I., & Robitzsch, A. (2015). Kindergartners' performance in two types of imaginary perspective-taking. *ZDM Mathematics Education, 47*(3), 345–362.

Van Hiele, P. M. (1986). *Structure and insight: A theory of mathematics education*. Orlando, FL: Academic Press.

Wilson, S. (1986). Feature frequency and the use of negative instances in a geometric task. *Journal for Research in Mathematics Education, 17*(2), 130–139.

World Health Organisation. (2006). WHO motor development study: Windows of achievement for six gross motor development milestones. *Acta Paediatr Suppl. Apr*; 450: 86–95.

Authors Biography

Dr. Aisling Leavy is a lecturer in Mathematics Education at Mary Immaculate College (MIC), University of Limerick, Ireland. Her research interests include children's mathematical thinking, the development of statistical reasoning, the mathematics preparation of pre-service teachers and use of teaching and research methodologies to support the development of conceptual understanding of mathematics in primary classrooms. Aisling is currently joint General Editor of Irish Educational Studies and Associate Editor of Statistics Educational Research Journal. She received her PhD from Arizona State University following completion of graduate and undergraduate studies at California State University Chico, St. Patrick's College Drumcondra and University College Dublin.

Dr. Jennifer Pope has been a lecturer in Early Childhood Care & Education in the Department of Reflective Pedagogy & Early Childhood Studies, Mary Immaculate College, Limerick since 2004. She also is the ECCE Student Placement Coordinator. She has a PhD in Paediatric Epidemiology and has a particular interest in promoting child well-being and the care and education of the under three age group. Jennifer's other professional interests include the enhancement of teaching and learning with early childhood undergraduate students, their experiences of practice placement, their professional development and the professional identity of ECCE graduates.

Deirdre Breatnach is a lecturer in the Department of Reflective Pedagogy and Early Childhood Studies at Mary Immaculate College (MIC), University of Limerick, Ireland. Deirdre worked at primary school level as a teaching principal, as a learning support teacher and as a class teacher up to 1999. She has been involved with the BA Early Childhood Care and Education programme in MIC since its establishment in 2003. Deirdre was a member of the research team who undertook a national study entitled 'An Examination of Concepts of School-readiness among Parents and Educators in Ireland' in 2014. Deirdre's research interests include language development in the early years, the transition from early years settings to primary school settings and immersion education within pre-school and primary school settings.

Chapter 8
Mathematizing Basic Addition

Allen Leung and Simon Hung

Abstract This chapter discusses teaching and learning of addition in early years under the mathematization framework. Addition-by-counting and the principle of Break-Apart-and-Make-Ten (BAMT) are used as examples to illustrate theoretical constructs in the process of mathematization: guided re-invention, horizontal and vertical mathematizations and the emergent modelling. The mathematizing process of addition is interpreted as progressive refinement of discourse mixing the gestural, verbal, iconic and symbolic registers. Subitizing is discussed under the interplay between mathematization and a materialist approach to mathematics.

Keywords Mathematizing · Addition · Counting · Subitizing
Guided reinvention · Early years of school

8.1 Introduction

This chapter presents a theoretical discussion on the teaching and learning of addition in early years and the role of mathematization in creating critical mathematical connections for a child between concept and self. The concept of addition is based on counting. Counting experiences and strategies are regular didactical activities in prior-to-school and junior primary mathematics classrooms. In the prior-to-school years, various counting concepts and skills are taught, for examples, verbal counting, object counting, counting out a group and counting in different arrangements (Clements & Sarama, 2009). In particular, addition-by-counting and Addition-by-BAMT (Break-Apart-and-Make-Ten) are two commonly used counting strategies taught. Mathematization is about designing pedagogical processes to turn concrete perceptual mathematical experiences gradually, from less precise to

A. Leung (✉) · S. Hung
Hong Kong Baptist University, Kowloon Tong, Hong Kong
e-mail: aylleung@hkbu.edu.hk

S. Hung
e-mail: cwhung@hkbu.edu.hk

V. Kinnear et al. (eds.), *Forging Connections in Early Mathematics Teaching and Learning*, Early Mathematics Learning and Development,
https://doi.org/10.1007/978-981-10-7153-9_8

more precise, into generalized mathematical concept reified by symbols and signs. This is followed by "playing" with the symbols and signs to gain further mathematical experiences. Counting-on and BAMT are two such mathematization processes to teach and learn basic addition. Different counting strategies converge to the same symbolic representation when critical invariant concepts are identified. This chapter discusses how to mathematize basic addition and to find conceptual connections between the counting-on process and the BAMT process.

8.2 Mathematization

Mathematics should emerge from and be close to children's experiences and thinking and should be learnt in an environment where pleasurable human activities are promoted and encouraged. This is the basic assumption of Freudenthal's theory of *mathematization* (Freudenthal, 1991). The central idea in mathematization is that mathematics is regarded as human activity connected to reality.

> ...mathematics as an activity of discovering and organizing in an interplay of content and form (p. 15)
>
> ... mathematics, unlike any other science, arises at an early stage of development in the then "common sense reality" and its language in the common language of everyday life. Why does it not continue in this way?" (p. 18)

Mathematics teaching and therefore learning must be relevant and realistic to children's everyday life situations. The term "realistic" refers to children experiencing problem situations which connect to their prior experiences. This means that problems presented to children can be a real-world situation, but that this is not always necessarily the case. Counting fictitious cartoon characters may be more relevant to children than counting apples if this provides a way of connecting that would not be possible otherwise.

In mathematization, the connections to be forged in teaching and learning are achieved through a process of *guided reinvention*.

> Children should repeat the learning process of mankind, not as it factually took place but rather as it would have done if people in the past had known a bit more of what we know now. (Freudenthal, 1991, p. 48)

In guided re-invention, children experience processes that mirror or replicate the processes by which mathematics was invented, but in a way that is accelerated due to advanced pedagogical settings. "Invention" refers to the learning process that achieves this replication, and "guided" refers to the pedagogical environment created to support the learning process. In such an environment, teachers inspire children to *re-invent* mathematical ideas, by developing "models" involving informal solutions and strategies that are *real* and relevant to the child. The children's collective solutions and strategies are then negotiated and generalized into more consolidated concepts which can be applied in a new mathematical situation where a solution is needed.

8.2.1 Horizontal and Vertical Mathematizations

In a pedagogical context, Treffers (1987) identified two types of mathematization that act to connect children's mathematical learning: horizontal mathematization and vertical mathematization.

> Horizontal mathematization leads from the world of life to the world of symbols. In the world of life one lives, acts (and suffers); in the other one symbols are shaped, reshaped, and manipulated, mechanically, comprehensively, reflectingly; this is vertical mathematisation. The world of life is what is experienced as reality, as is symbol world with regard to its abstraction. (Freudenthal, 1991, p. 42)

In horizontal mathematization, children develop contextual tools (in the broadest sense) to help them to solve problems in real-life situation. The models children develop to do this enable them to transfer the problem situation into a mathematical one. Horizontal mathematization is therefore a process in which children's contextual solutions, as models, are gradually transformed into generalized mathematical concepts. Examples of activities conducive to this transformation are as follows: describing the specific mathematics in a context, visualizing a problem in different ways, discovering relations and patterns and translating a real-world problem into a mathematical problem.

When children focus on generalized mathematical concepts and use them to build models to solve problems within the mathematical system itself, they enter into the process of vertical mathematization. In vertical mathematization, symbols play a dominant mediation role, helping children to make sense of their world. Activity examples in vertical mathematization are as follows: representing a relation symbolically, refining solution models, using different models, combining and integrating models. In pedagogical practices with children, the boundary between the real world and the mathematical world is not well-defined. When children learn addition-by-counting, they have to use reified number symbols. Reified number symbols are metaphorically an ontological mathematical object that is *objectified* through a process of procedural action and internalization (Sfard, 1994). Children use number symbols and manipulate them in some way (vertical mathematization) to come up with a counting model using fingers (horizontal mathematization). Consequently, the difference between horizontal and vertical mathematization is not necessarily clearly defined or definable, and they become progressively symbiotic with each other.

8.2.2 Didactical Phenomenology

The concepts of horizontal and vertical mathematization provide a basis for the phenomenological exploration or the use of the context, which is a core characteristic of mathematization. Figure 8.1 depicts a cyclical mathematization process where horizontal mathematization and vertical mathematization intertwine in reflection and application.

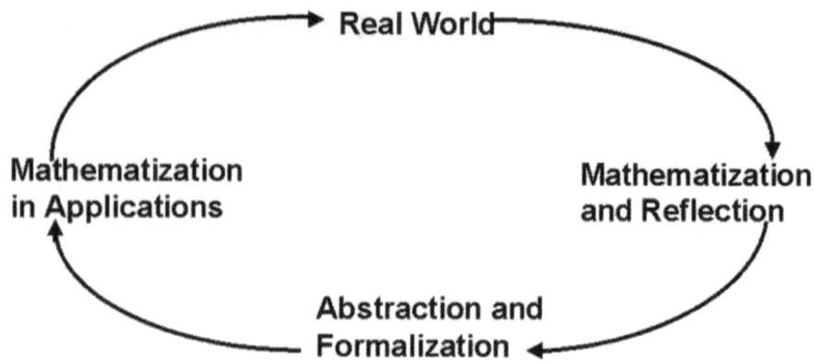

Fig. 8.1 Concept and applied mathematization (de Lange, 1996)

The phenomena in the selected real situation or contextual problem, where the mathematical concepts are embedded, should be the source of concept formation for children. Children then explore the phenomena through horizontal mathematization activities in order to identify the mathematical aspects of the problem, to discover patterns and relations and to form informal solution models for the problems. Through refining solution models, children enter into vertical mathematization to develop formal symbolic mathematical concepts. These concepts are then applied to new real problems, thus forming connections, reinforcing and strengthening the concepts. This can be illustrated through two approaches to addition.

8.3 Mathematizing Basic Addition

One can interpret the "horizontal → vertical" mathematization process as a kind of "mathematics reification" process where contextualized mathematics evolves to form connections with symbolic-based mathematics found in more formal mathematical conventions. In horizontal mathematization, children use previously reified mathematical objects (e.g. the number symbols) to develop contextual models and solutions to connect to a new reified concept (e.g. addition). The new reified object can then be used in a new mathematization process (horizontal or vertical), so on and so forth. In this way, a horizontal → vertical mathematization process *collapses* into a mathematical object once reification takes place. This accumulative epistemic progression is the basis for guided reinvention as a pedagogical process.

8.3.1 Mathematization with Addition-by-Counting

Counting-on serves as a core strategy in the learning trajectory towards mastery of basic addition. Addition-by-counting requires an understanding of cardinality as children learn that the name of the group is the last number word spoken after counting every group member. In counting-on, children learn to count from any number onwards without relying on the verbal sequence starting from 1. Addition-by-counting can appear to be deceptively simple and intuitive; however, the whole procedure requires children to have developed concentrated coordination of multiple counting and a mastery of counting-on, and these and the procedural process create a significant cognitive load for children. Based on the concepts and skills developed when using counting-on strategies, beginning addition can be performed as addition-by-counting.

An example of addition-by-counting introduces the process and key concepts examined in this chapter. When children perform the addition 6 + 7, the following action sequence may take place:

1. Put the number 7 *in your mind*
2. Raise six fingers for the number 6
3. Count on from the number *in your mind* using the raised fingers one by one
4. Get the answer when all the raised fingers are counted

Figure 8.2 illustrates what can happen when an addition-by-counting strategy is used. One starts and ends with a *reified number* (initially 7 and finally thirteen 13).

Before learning addition-by-counting, children will need to have conceptual understanding of both the numerals 7 and 13 as symbols representing objects as quantity that are called "number". In the process of adding on, an action process occurs that combines finger movement, verbally counting-on, and symbolic numeral representation with concrete physical objects (the six fingers). In this example, the abstract symbol 6 does not explicitly appear in the counting process and is replaced by contextual physical objects, although a child may visualize a numeral during this process. This process works to "mathematize" the problem situation, that is, to bring about the emergence and use of an adding algorithm as a problem-solving strategy. Figure 8.2 can therefore be interpreted as representing a finger-motion-concept *assemblage* (de Freitas & Sinclair, 2013), embodied in action. The assemblage is germane for children to comprehend and integrate the concepts and skills of addition. The integration occurs gradually when this assemblage "grows" from routine movements to actualization where children internalize meaningful mathematical understanding.

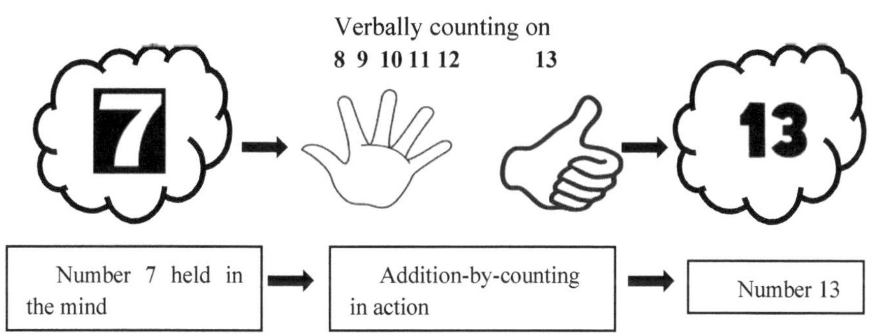

Fig. 8.2 In addition-by-counting, a reified number is blended with gesture and verbally counting to reach another reified number

Fig. 8.3 Children associate objects with reified symbols

When a child performs object counting for six Hello Kitties and two Hello Kitties for example, objects that are both realistic to the child and formal mathematical symbols are involved. Suppose the child has preliminary understanding of number symbols and is given six Hello Kitties in a group and two Hello Kitties in a group. Preliminary here means that child can intuitively associate these symbols with verbal counting and quantity through previous horizontal → vertical mathematization processes. With these reified symbols, children associate the six Hello Kitties with "6" and the two Hello Kitties with "2" (Fig. 8.3).

Pedagogically engaging the guided re-invention approach, children can be motivated to put these two groups of Hello Kitties into one group of Hello Kitties (Fig. 8.4 is a possible grouping) and to associate this new group with a number symbol by counting the objects. After deliberation, the symbol 8 now should have a new meaning for the children, that is, "8 Hello Kitties comes from putting six Hello Kitties and two Hello Kitties together in a group". This can be seen as a horizontal mathematization process.

At this point, the children can be guided via further activities to gradually progress towards vertical mathematization:

> "Putting 6 Hello Kitties and 2 Hello Kitties together in a group and count to get 8 Hello Kitties"
>
> "6 and 2 put together is 8"
>
> "6 and 2 more is 8"
>
> "6 + 2 = 8"

When children are engaged in vertical mathematization, mathematical objects (in this case number symbols) take over and daily life objects (like the Hello Kitties) lose their dominance. In the process of conceptualizing the mathematical

Fig. 8.4 Eight Hello Kitties comes from putting six Hello Kitties and two Hello Kitties together in a group, the first step in mathematizing basic addition

knowledge, "6 and 2 is 8" and "6 + 2 = 8" become objects that children can use to connect to and acquire further mathematical knowledge. The purpose of guided reinvention then is to reify prior mathematical experiences and to make connection between different levels of learning experiences lead to vertical mathematization. Making sense of a mathematical experience is an important motivation for children to learn. Children's choices of how to order the spatial positions of the two groups of Hello Kitties can be used to open up an awareness of the commutative property of number addition. That is, counting the group of two Hello Kitties first follows by counting the group of six Hello Kitties or vice versa gives the same result: eight Hello Kitties. This sense making can be further developed through a mathematizing process to reach the equivalent relationship "6 + 2 = 2 + 6". With a sensible reification, that is, construction of mathematical objects that make sense to the children, mathematization processes can connect with each other, enabling children to construct personally significant mathematical conceptual knowledge and skills.

Guided reinvention in this sense serves as an overarching pedagogical design heuristic that enables children to take initiatives to generate their own knowledge under deliberately designed experiences in the learning environment. In the process of guided reinvention, teachers design learning experiences for children that require symbolizing, that is, children need to find for themselves progressive layers of mathematical symbols that will enable connections to be made from personal to formal symbols. Daily life objects that are real to the children (e.g. Hello Kitties) are linked to known symbols (e.g. numerals). The act of putting things together leads to the need to make symbols for addition.

8.3.2 Mathematization with Addition-by-BAMT

Building on counting-on as a strategy, addition can be interpreted through the subitizing approach Break-Apart-and-Make-Ten (BAMT) (Clements & Sarama, 2009), shown here using the addition of six and seven using the following steps:

1. Break apart 7 into 4 and 3
2. Leave 3 behind
3. Combine 4 and 6, count to get 10
4. Bring 3 forward and combine it with 10, count to get 13

Apart from the foundational skill of counting-on which is based on the concept of cardinality, Addition-by-BAMT both requires and develops the concepts and skills of composition and decomposition of numbers. An interesting and important observation for Addition-by-BAMT is that, in the example 6 + 7, children will usually first breaking apart 7 into 4 and 3 to make 10 by combining 4 with 6, but seldom breaking apart 6 first to combine 3 with the 7 to make 10. Reading from left to right (a linguistic dimension of mathematics), children typically see the numeral "6" first and look for an addend to make 10, thus breaking apart the 7.

Addition-by-BAMT therefore requires making choices about which number to "break apart" to make 10. Three composition/decomposition actions are involved in this: (1) mentally compose an imaginary "missing addend" to make 10 (2) decomposing a number in the addition to find the "missing addend" and (3) compose to make 10 and see what is left. As complicated as it looks conceptually, the whole addition process is usually achieved quickly once the missing addend is identified.

Addition-by-BAMT is less demanding in procedural steps than addition-by-counting; however, this is at the expense of a more demanding conceptual understanding, that is, the need to find a missing number. In contrast, addition-by-counting is conceptually more demanding in the counting procedures, as it is based on intuitive concepts of cardinality and one-one counting principle. Thus, the critical distinguishing feature in Addition-by-BAMT is determining how to find the missing addend.

The 5-Frame pictorial representation of 10 is a tool that illustrates mathematizing the missing addend feature in Addition-by-BAMT. In Fig. 8.5, 10 as a whole is represented as a rectangular frame partitioned into two rows of five equal squares.

When putting less than 10 items in the squares (one in each), the number of unoccupied squares *can be seen visually* as the "missing number". A story can be used to start the situational discourse for 6 + 7, such as in the following example.

"In a zoo there are two Panda Houses each contains 10 rooms. 6 Big Pandas lives in House A and 7 Little Pandas live in House B (Fig. 8.6). *The visitors are not happy about this arrangement because they want to see a full Panda House with Big Pandas and Little Pandas. If you are the zoo keeper and you have to move some Pandas around to fill a Panda House to make the visitors happy, how would you do it? How many more pandas would the zoo need to have to have 2 full panda houses? How many more Pandas to reach a second '10'?*

Children can start to play with this situation presented in the story by moving the Big Pandas and the Little Pandas around to create either a full House A or a full House B and to count how many more Pandas would be needed to have House B full (reach a second "10") and to ask children to explain the way a solution was developed. The child may say something like this:

1. Notice there are four *free rooms* in House A with six Big Pandas
2. Take four Little Pandas from House B with seven Little Pandas
3. Put the four Little Pandas one each into the four *free rooms* in House A

Fig. 8.5 A 5-frame representation of 10

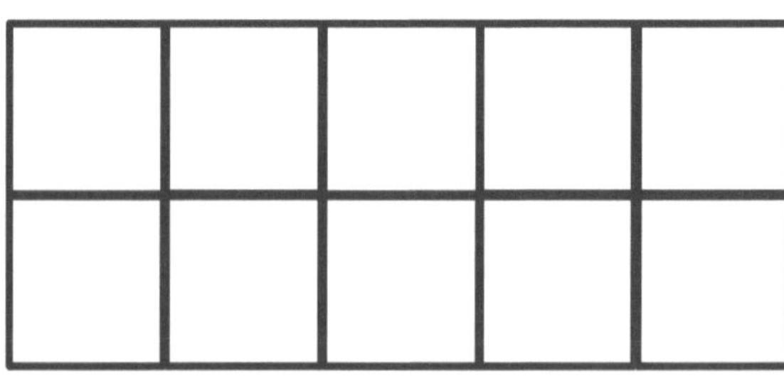

Fig. 8.6 The Pandas story

4. Now House A is full of Pandas
5. There are three Little Pandas left in House B
6. There are all together one full House A with 10 Pandas and a House B that has only three Pandas
7. Seven more Pandas would be needed to fill House B (Fig. 8.7)

Of course, children could have chosen to fill House B instead, and the number of free room now becomes three, and the procedure would be generically the same. The children now have developed a model of "adding Pandas" using the 5-Frame. Based on these possible discourses, the teacher can develop different stories using the 5-Frame metaphor and encourage children to find ways to relate the empty squares to the whole rectangular frame when solving addition problems. The different solutions children create to solve the problem can be generalized to become a model for addition in mathematical contexts. Addition-by-BAMT, in contrast to addition-by-counting, requires a representation of numbers adopting subitizing strategy, in particular, conceptual subitizing (Clements, 1999; Clements & Sarama, 2009). Under the 5-Frame, the concept of cardinality is accentuated under different possible pattern arrangements when implementing conceptual subitizing. Numerical relationships arise from pattern combinations that give children a solid learning approach to addition.

When progressing from horizontal mathematization to vertical mathematization, transitionary discourse composing of symbols, pictures and wordings could occur bridging to a formal symbolic discourse. Figure 8.8 can be interpreted as a transition discourse for the formal expression $6 + 7 = 6 + 4 + 3 = 10 + 3$.

Fig. 8.7 A result of Addition-by-BAMT for the Pandas story

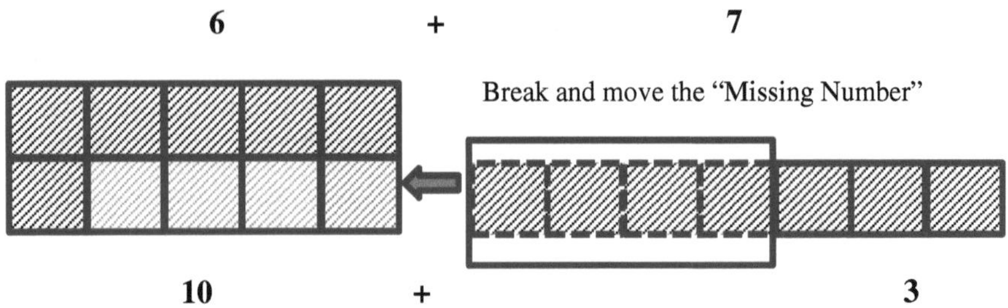

Fig. 8.8 A transition discourse for 6 + 7 = 6 + 4 + 3 = 10 + 3

Iconic, verbal and symbolic registers are mixed to visualize the meaning of the addition process which could create a mathematical reality for children in terms of the Addition-by-BAMT metaphor. This brings us to emergent modelling in mathematization.

8.4 Emergent Modelling

In the process of mathematization, Gravemeijer (1994) identified four levels of mathematization that provide a perspective on the progression of mathematical symbols and how they emerge and reshape. The four levels are as follows: situational (within a context), referential (model of), general (model for) and formal (conventional mathematical notation). Mathematical concepts and symbols emerge in the situational level from activities that take place in realistic situations. "Realistic" is what can be "imagined as real" that may differ from objective reality. At this level, children's verbal expressions and gestures form symbols, which develop into informal contextual models to solve mathematical problems. "Putting six Hello Kitties and two Hello Kitties together in a group and count to get eight Hello Kitties", "Take four Little Pandas from House B with seven Little Pandas and Put the four Little Pandas one each into the four free rooms in House A" are verbal descriptions of a bodily action on physical objects to solve a counting problem.

At the referential level, children can be guided to develop more symbol-based expressions that act as a solution model of a similar situation. "6 and 2 putting together is 8", "break apart 7 into 4 and 3, leave 3 behind, combine 4 and 6, count to get 10, bring 3 forward and combine it with 10, count to get 13" describe an action counting procedures, when putting "things" together. This emergent model of symbol development condenses and replicates the process of how mathematical symbols were historically invented and developed. In this respect, mathematization through guided re-invention invites children to develop their own reification of mathematical ideas, while participating in activities that mirror human endeavour to solve realistic problems. From a learning perspective, this emergent model opens up opportunities for children to make discerning mathematical decisions through the production of different representations of an idea (e.g. addition) via a progressive development of mediating mathematical symbols. Experiencing differences is a powerful way to learn. Marton's Theory of Variation asserted that:

> no discernment (of features or aspects) can happen without the experience of difference. But no difference can be experienced without the simultaneous experience of the things that differ. And two things cannot be experienced simultaneously – as two things – without being discerned. (Marton, 2015, p. 66).

The simultaneous presence of the different symbol-making discourses enables children to make sense of a mathematical concept from different points of view. In learning addition as discussed above, children experience addition through physically putting things together, verbally describing what is happening, transferring of experience to general situation and manipulation of mathematical symbols. These combined experiences form the basis of the concept of addition, Thus, the four levels of mathematization can be regarded as four different types of mathematical discourses for children to learn mathematics, not as hierarchical levels, but rather as four interconnected discernment experiences, integrated into a whole mathematical experience: situational discourse (symbolizing within a situation), referential discourse (symbolizing about a situation), general discourse (symbolizing about a type of situation) and formal discourse (symbolizing without a situation). These four discourses stand under a mathematical concept and thus become the foundation to understand a mathematical concept. The transitionary discourse suggested in Fig. 8.8 can be interpreted as an integrated discourse for children to discern the meaning of addition.

8.5 Subitizing

Subitizing, the cognitive ability which children usually use to develop their number sense and arithmetic skills, could be deliberated under the process of mathematization. The basic meaning of subitizing is to recognize numbers without using any mathematical process. For example, a child can "see 3" when he/she encounters three dots in certain spatial arrangement without using any prior learned knowledge. This innate ability is called perceptual subitizing. An important mathematical precept that perceptual subitizing entails is making units to count. For example, certain three-dot pattern can be a "thing" for a child to use as a (counting) unit, and this thing is connected to a unique number word "three". In a sense, perceptual subitizing can be regarded as a kind of reification, when a piece of experience is detached from the whole. However, "cutting out" pieces of experience, keeping them separate and coordinating them with number words are no small task for children. (Clements, 1999, p. 401). Perceptual units can be seen as "pseudo-symbolic" objects, in the sense that they can be manipulated as mental entities that could gradually transform into reified formal symbols under a mathematization process. The perceptual subitizing counting units can be combined to form other spatial patterns which may progress to become other counting units.

This combination process is called conceptual subitizing (Clements, 1999). It is an advanced-organization cognitive activity where part–whole relationships are discerned among different spatial patterns.

Penner-Wilger et al. (2007) studied children's foundations of numeracy via subitizing, finger gnosia and fine motor ability. These abilities were found to support the development of early childhood mathematical skills. Doane (2014) studied the effect of teaching subitizing and addition skills to kindergarten students with mathematics deficits using curriculum materials based on 5-Frame and 10-Frame. The study produced evidences to support claims that children can be taught to subitize through direct instruction (Sousa, 2008). Subitizing is closely related to children's spatial structuring ability. van Nes (2009) identified four consecutive phases in the development of young children's spatial structuring ability: unitary, recognition, usage and application. These phases describe how a child progresses gradually from beginning to recognize fundamental spatial structures like simple dot patterns (perceptual subitizing), to use and apply these structures to abbreviate numerical procedures like basic addition (conceptual subitizing). Furthermore, these phases were compared with the learning levels in emergent modelling, harbouring a mathematization outlook for subitizing.

Mathematics as a human activity is the main tenet of Freudenthal's theory of mathematization. Activity involves human action; therefore, mathematics concept development should be corporate with what we do. In proposing a new materialist ontology in mathematics education, de Freitas and Sinclair (2013) expounded Gilles Chatelet's, a philosopher of mathematics, approach to mathematics in which mathematical activity "involves both actualizing the virtual and realizing the possible" (p. 461). In a mathematical activity, e.g. children counting or adding, realizing the possible is to play by the rules of logic when children strategize the "rules of the game", e.g. counting-on, to carry out activity tasks. The virtual refers to:

> that which is latent in matter—but what is latent is not essence, ideal or form, but rather mobility, vibration, potentiality and indeterminacy. Mathematical activity taps this potential (primarily through diagramming and gesture) and engenders new objects of inquiry. (de Freitas and Sinclair, 2013, p. 462)

Actualizing the virtual is a creative act where physical materials and mathematical ideas are entangled in an object of enquiry, for example, the 5-Frame for basic addition discussed in the previous section. A child uses the 5-Frame as a source of action to perform addition task. The virtual points to the indeterminacy of the source of action as the 5-Frame carries potential of action rather than prescribed arithmetic procedure. A child uses the 5-Frame in a way that suits his/her proclivity at the moment but such a way may change as the learning progress. The virtual is dynamic with uncertainty while the possible relates to agreement with conventional logical rules and perceptual habits. In this approach, mathematics is seen as an entangled human-material activity where a learner realizes and actualizes, respectively, possibility and virtuality embodied in matters which may eventually give rise to meaningful mathematical concepts.

With this materialistic approach, mathematizing basic addition can be interpreted as a dynamic evolving assemblage in action:

Child---Gesturing/Diagramming---Basic Addition

where perceptual subitizing is gradually transformed into conceptual subitizing. Moreover, perceptual subitizing and conceptual subitizing may eventually merge into one interactive process. In this case, the boundary between horizontal mathematization and vertical mathematization may not be clearly defined as the assemblage does not decontextualize basic addition as an entity devoid of material experiences; rather basic addition is a child's action both on a virtual material object of inquiry (for example, finger gesture, the 5-Framd or 10-Frame diagram) and a possible concept of addition (for example, the conventional symbolic arithmetic). Figure 8.9 depicts a "three-in-one" nature of the assemblage where the virtual and the possible coexist under the child's action.

The process of mathematization for basic addition in the light of this materialistic approach takes on a more recursive orientation. Children's creative action of strategizing addition using realistic (materialistic) artefacts is as mathematically meaningful as teaching them how to use formal symbolic representation. Thus, horizontal mathematization and vertical mathematization may sometime be better understood as two parallel processes rather than merely a unidirectional "reification" process focusing on how to turn children's realistic experiences into formal symbolic manipulation. Developing children's ability to actualize the flexible usages of the 10-Frame in understanding addition is as pedagogically relevant as

Child	**Virtual Material Object of Inquiry**	**Possible Mathematical Concept**
Actor	Gesturing and Diagramming	Basic Addition
Subitizing using finger gestures and dot patterns		Possible Symbolic Logic 5 + 1 = 6
Subitizing using the 10-Frame	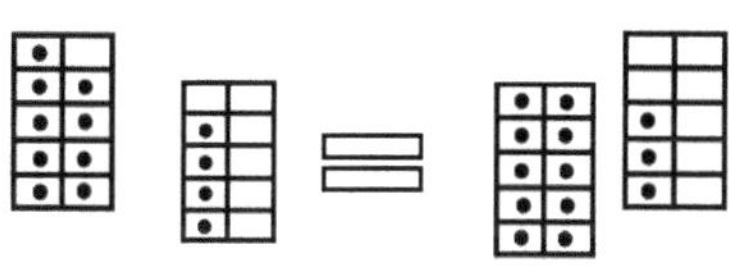	Possible Symbolic Logic 9 + 4 =13

Fig. 8.9 The three-in-one nature of the Child—Gesturing/Diagramming—Basic Addition assemblage

teaching children how to do symbolic addition. These two pedagogical processes engender each other, and they should go hand in hand. In the same way, perceptual subitizing should not be regarded as more primitive than conceptual subitizing as perceptual subitizing can be cultivated from conceptual subitizing. Studying the intra-dynamic of an assemblage may shed new light on the mathematization process.

8.6 Concluding Remarks

Engaging children in mathematical experiences with a view to seeing the mathematization that is possible and that occurs when children find solutions to problems highlight connection possibilities. This is because mathematization stresses the importance of children producing their own (knowledge) construction, and through interactivity finding ways to connect across and to different mathematical concepts. These characteristics of mathematizing activities are core to pedagogical practices that support re-inventing (historically) the development of mathematical concepts. As a human activity, "adding things together" takes on various forms and interpretations that have emerged from diverse socio-cultural origins where creative tools and algorithms were employed. This development reflects the basic assumption of mathematization that mathematics is an activity of discovering and organising the interplay of content and form (Freudenthal, 1991).

Properties of addition (for examples, commutativity, associativity, identity, units) form a foundation to build advanced mathematical structures in algebra and geometry. These can be interpreted using metaphors such as combinations, extension of length and motion along a straight line. Finding ways for children to "mathematize" these metaphors in children's early years can sensitize children to the interplay between concept and form, and support children to connect mathematical experiences with personal knowledge, preparing them for their future in the ever developing mathematical world.

The materialist ontological approach discussed in the subitizing section suggests a pedagogical orientation where human activities, artefacts/tools and mathematical concepts are entangled together into a whole. This implies that horizontal mathematization and vertical mathematization should not be treated as separated processes, rather they should become part of each other. A pedagogical implication for this is when teaching basic addition, teachers should design activities for children to construct material objects to, for example, illustrate the meaning of $2 + 3 = 5$, instead of the usual practice where teachers provide the artefacts as a transitional mean to reach the symbolic expression. This, of course, does not happen at the beginning of the mathematization process, but rather at moments when children are introduced to formal arithmetic operation. Hence, actualizing the virtual can be thought of as situating an abstract symbolic expression like 2 + 3 = 5 in a realistic context that involves motion, openness and flexible interpretation. This kind of pedagogical approach in early years mathematics education will be an interesting and novice research direction that could open up a new discussion on mathematizing basic addition.

References

Clements, D. H. (1999). Subitizing: What is it? Why teach it? *Teaching Children Mathematics, 5*, 400–405.

Clements, D. H., & Sarama, J. (2009). *Learning and Teaching early math: The learning trajectories approach*. New York: Routledge.

de Freitas, E., & Sinclair, N. (2013). New materialist ontologies in mathematics education: The body in/of mathematics. *Educational Studies in Mathematics, 83*, 453–470.

De Lange, J. (1996). Using and applying mathematics in education. In A. J. Bishop et al. (Eds.), *International handbook of mathematics education, Part one* (pp. 49–97). Kluwer Academic Publisher.

Doane, T. L. (2014). *The effects of teaching subitizing and addition skills of kindergarten students with math deficits (Unpublished Doctor of Education Thesis)*. Baltimore, MD: Johns Hopkins University.

Freudenthal, H. (1991). *Revisiting mathematics education. China lectures*. Dordrecht: Kluwer Academic Publishers.

Gravemeijer, K. P. E. (1994). *Developing realistic mathematics education*. Utrecht: CD-ß Press/Freudenthal Institute.

Marton, F. (2015). *Necessary conditions of learning*. New York and London: Routledge.

Penner-Wilger, M., Fast, L., LeFevre, J., Smith-Chant, B. L., Skwarchuk, S., & Kamawar, D., et al. (2007). The foundations of numeracy: Subitizing, finger gnosia, and fine motor ability. In *Proceedings of the 29th Annual Conference of the Cognitive Science Society* (pp. 1385–1390). Mahwah, NJ: Erlbaum.

Sfard, A., & Linchevski, L. (1994). The gains and pitfalls of reification. *Educational Studies in Mathematics, 26*, 191–228.

Sousa, D. (2008). *How the brain learns mathematics*. Thousand Oaks: Corwin Press.

Treffers, A. (1987). *Three dimensions, a model of goal and theory description in mathematics education*. Dordrecht: Reidel.

van Nes, F. T. (2009). *Young children's spatial structuring ability and emerging number sense* (unpublished doctoral thesis). FIsme, Universiteit Utrecht, Utrecht, the Netherland.

Authors Biography

Dr. Allen Leung is Associate Professor in Mathematics Education at Hong Kong Baptist University. Allen's research interests include task design in the mathematic classroom under a tool-based environment, reasoning and argumentation in dynamic geometry environment, development and application of variation theory in mathematics teaching and learning and STEM education. Allen is currently leading a project working with school teachers exploring tool-based task design in the mathematics classroom leading to the development of an inquiry-based modelling approach in STEM education. Allen co-edited the book Digital Technologies in Designing Mathematics Education Tasks for the Springer Book Series Mathematics Education in the Digital Era.

Mr. Hung Chun Wah is now a senior lecturer in Hong Kong Baptist University dedicated to Mathematics curriculum and instruction and to provide education training for teachers and parents. His recent publications include series of activity books for kindergartens "Early Childhood Mathematics in Focus", articulating to the primary and secondary series of "Mathematics in Focus". He has been collaborating with kindergartens for years on the projects about early childhood mathematizing experiences.

Chapter 9
Connecting the Mathematics Identity of Early Childhood Educators to Classroom Experiences for Young Children

Sandra M. Linder and Amber M. Simpson

Abstract This chapter examines the mathematics identities of early childhood educators in birth to five-year-old settings. Connections are explored between how these educators perceive their role in promoting mathematics in child care settings and the opportunities afforded to the children in these settings. Results are presented from a phenomenological investigation of how participants view teaching and learning mathematics and how they experience mathematics teaching with young children. Participants (n = 8) were Head Start and Early Head Start teachers involved in a year-long mathematics professional development experience. They were chosen based on their level of implementation of practices recommended to them through professional development. In-depth interviews and classroom observations were conducted and results compared to determine how mathematics identity influenced practice. Example vignettes describe how these teachers implemented mathematics with young children. Embedded in these examples is a discussion of next steps for these teachers in terms of their own professional growth.

Keywords Identity · Professional development · Mathematics
Phenomenology · Prior-to-school

S. M. Linder (✉)
Clemson University, Clemson, USA
e-mail: sandram@g.clemson.edu

A. M. Simpson
Indiana University, Bloomington, Indiana, USA
e-mail: ambmsimp@iu.edu

V. Kinnear et al. (eds.), *Forging Connections in Early Mathematics Teaching and Learning*, Early Mathematics Learning and Development,
https://doi.org/10.1007/978-981-10-7153-9_9

9.1 Introduction

When I first started teaching

Math…

It was not my strong point

I didn't like it, know it

I wasn't capable of teaching it

I stifled myself and the children

Now

Math…

It is not as intimidating

I have fun

I am open, no blinders

I have grown

The above-found poem (Butler-Kisber, 2010) is a representation from an early childhood educator in our study. It was constructed by the second author to illustrate and reinforce how identity, in general, is a complex, multifaceted, and dynamic construct; yet, individualistic by nature (Simpson, 2015; Boaler, 2002) and subject-specific (Nasir, 2002; Spillane, 2000). In essence, identity is 'at the core' of the teaching profession. It provides a framework for teachers to construct their own ideas of 'how to be', 'how to act', and 'how to understand' their work and their place in society" (Sachs, 2005, p. 15). Specific to this chapter, we examine how one perceives and narrates who they are as an early childhood educator in relation to mathematics, which has been shown to have profound effects on the classroom environment and the learning opportunities afforded to students from instructional decisions (e.g., Chronaki & Matos, 2014; Drake et al., 2001) in addition to teachers' ability to foster positive mathematics experiences for children (Gujarati, 2013). Further, some teachers of young children have to redefine their own mathematics identities that have been shaped by negative experiences with mathematics as students (Drake et al., 2001; Gunderson et al., 2012).

Transformative professional development (Borko, 2004; Smith, 2001; Sztajn et al., 2007) or experiences that attempt to change beliefs and alter instructional practices of educators to align with best practices in mathematics education not only require extensive time commitments on the parts of both the facilitator and the participant but also require that the facilitator understands the participant in terms of their classroom context and how they perceive their role within that context. Central to transformative professional development experiences is the need to make content relevant and meaningful for participants by providing explicit connections to their own classroom contexts (Margolinas et al., 2005; Norton & McCloskey, 2008). Understanding participants' mathematics identity, or as Sachs (2005) describes, 'how they act,' and 'how they understand' their role as a mathematics educator in an early childhood setting is a critical component for making these connections. As the opening poem exemplifies, transformative professional development has the potential to serve as a catalyst for a positive shift in how an early childhood educator views their role in the teaching and learning of mathematics.

9.2 Navigating Professional Development

The need to better understand child care teachers' mathematics identity emerged through an investigation of ongoing professional development focused on early childhood mathematics. The project, *B*uilding *E*nvironments for *E*arly *M*athematics *S*uccess (Project BEEMS), funded through the PNC Foundation, called for the implementation and evaluation of a sixteen-hour professional development symposium (split between two full-day meetings in October 2013 and January 2014) and ongoing follow up support geared toward encouraging child care teachers and teacher assistants from Head Start and Early Head Start settings to create classroom environments that fostered early mathematics success for children from birth to age five. Head Start is a federally funded program in the USA, originally implemented in 1965 and designed for children ages three to five from low income and impoverished backgrounds to improve school readiness. Early Head Start was implemented as an arm of the Head Start program in 1994 and serves pregnant women, infants, and toddlers through educational initiatives for parents and child care services for children. At present, Head Start programming has supported over 30 million children (Office of the Administration for Children and Families Early Childhood Learning and Knowledge Center (ECLKC), 2016). Project BEEMS influenced the mathematics environment of twelve Head Start and Early Head Start child care centers (n = 90 teachers and teacher assistants from 40 classrooms across the twelve sites).

The Project BEEMS professional development was designed to provide participants with experiences that are grounded in best practices in mathematics education, including the use of collaborative opportunities, hands-on and engaging tasks, context-specific examples, and opportunities for reflection (Cobb, 2000; Gellert, 2008; National Council of Teachers of Mathematics, 2000; National Research Council, 2001). Participants learned how to structure and support early childhood mathematical environments and how to use this environment to build understandings in five specific mathematics content areas: numbers and operations, geometry, measurement, algebra, and data analysis. Participants worked to build understandings about how to encourage young children to develop the processes necessary to engage in mathematical experiences. These processes build upon children's natural desire to question, explore, and discover new knowledge about the world around them. In between and following formal sessions, we visited each classroom location monthly and worked with the participants by providing model lessons, observing and providing feedback, teaching concurrently with participants, or meeting as a group following instruction.

Within the context of these ongoing visits, we began to notice discrepancies in how participants enacted the mathematics practices they learned during professional development sessions. For some, engaging in more and deeper levels of mathematical play and mathematics tasks throughout the day was an easy change. For others, it was more of a challenge. When discussing with teachers how they felt about the changes they were making, often teachers related the process of change to their own conceptions about mathematics and mathematics instruction; that is, 'I didn't think the children would be able to do this' or 'I never learned math this way'. Understanding these conceptions became integral to navigating professional development pathways. Next steps for each participant were determined based on these discussions. For example, when they voiced feelings of fear or inadequacy in trying out mathematical tasks during small group time, the project teams offered to model and co-teach with participants to increase levels of confidence.

9.3 Looker Deeper at Identity

Following year one professional development implementation, we reflected on the overall process and the changes we noticed for each participant. When discussing their pathways of professional development, we became increasingly interested in how participants connected their own conceptions about themselves and their role as mathematics educators to their instructional practice within the classroom context. We decided to examine how a subsample of participants perceived their own identity in regards to early childhood mathematics instruction and compare these perceptions to videos of their instructional practice (either during circle time, free play during centers, or within a small group task). The research question guiding this exploration was: How do Head Start and Early Head Start teachers perceive themselves and their role during mathematics experiences in child care environments and how do these perceptions compare to enacted practice?

Maximal variation sampling was used to select eight participants from the overall 90 participants that were engaged in ongoing professional development that varied in terms of years of experience, age of students, and center location. We included participants from both Early Head Start and Head Start settings, and we also focused on teachers who were open to ongoing video data collection of their classroom practice, which was not the case for all participants. We wanted to better understand how teachers who made a substantive commitment to addressing their mathematical practice perceived themselves as mathematics educators and thereby better understand how to approach next steps in professional development.

A phenomenological approach was used to understanding these teachers' mathematics identity by conducting in-depth, semi-structured interviews. Each interview lasted approximately 40–45 min and was typically conducted at the child care center following dismissal. One interview was conducted via telephone because the participant had transferred to a different state prior to data collection. Interviews focused on how teachers viewed themselves and their practice related to

Table 9.1 Interview questions

• Tell me about a typical day in your classroom
• What are your children's favorite things to do throughout the day? • How would you describe your role as the teacher throughout the day? During circle time? Small group time? Center time? • What are important things for these kids to know or be able to do prior to entering kindergarten? • When I say mathematics, what does that word mean to you? • How are you incorporating math throughout the day? • Is it important for young children to understand mathematics? Why or why not? • How should young children experience/engage in mathematics? • Can you describe an example of a mathematical task you have implemented? • How would you describe a strong early childhood teacher? And one that is not as strong? • What was experience with mathematics as a child? How did you experience mathematics in school (or as student)?

early mathematics. The interview questions are listed in Table 9.1, and each interview was transcribed verbatim and read through multiple times to gain an overall understanding of the data.

A thematic coding approach, common to transcendental phenomenological design (Moustakas, 1994; Moerer-Urdahl & Creswell, 2004), was used to examine each interview transcript. Each participant interview was analyzed individually to extract essential units of meaning (Moustakas, 1994) that were then clustered into overarching themes. These themes were then examined across the participants to determine commonalities. Although we examined individual participants, we also began to notice consistencies among the group. Below we present the common characteristics for the group of eight participants. In addition, we include portrayals of two participants' mathematics identity as compared to their instructional practice and how this analysis could inform future pathways of mathematics professional development.

9.4 Results

Three overarching themes emerged across participants from an analysis of these eight teachers and how they described themselves as early childhood mathematics educators: (1) respect for children; (2) feelings of inadequacy; (3) goals for instruction. These themes are detailed below.

9.4.1 Respect for Children

Participants described how important it was for them to get to know the children in their classroom, both in terms of their knowledge and in terms of their personal contexts. When describing their primary roles within the classroom, participants used terms such as 'understanding and meeting children's needs,' 'listening to children's conversations,' and 'knowing their home environments.' These descriptors indicate that participants' care for and value children's experiences both inside and outside of the classroom.

> I can learn if they are used to certain vegetables we have or certain meats. I can learn if they sit down family style at home with their families to eat their meals. I pick up on a lot of self-help skills because during that time they have to set the table. And then after the meal, they have to clean up. And so I think I learn, I look after self-help skills and how they relate to each other during meal time. If they eat family style at home with the parents. (Interview 2, p. 1)

In addition, participants respected children's capability and responsibility for their own learning. They often described their primary role as facilitating and encouraging children by asking questions as they engaged in play, as well as allowing children to have autonomy and direction over tasks with the teacher guiding or scaffolding when necessary. This finding was particularly insightful, as it was a direct focus in the Project BEEMS professional development and is consistent with best practices in early childhood (e.g., National Association for the Education of Young Children, 2002) and mathematics education (e.g., NRC, 2001). Participants further indicated that children learned better from each other than from direct instruction from the teacher.

> My favorite part of the day is work time because we are able to move around in the classroom and observe the children in their natural state sort of speak. Where they are comfortable in their play, and then they are interacting with their peers. And we get to see certain skill levels come out that we wouldn't ordinarily see if we were sitting with them, talking about something, and trying to get them to do this skill. We can interact with them and talk with them and get to know a little bit more about them. And they get comfortable with us so that when we are together in a group setting, they are more apt to participate, more willing to feel comfortable in giving answers because they have a relationship with us. (Interview 3, p. 1)

9.4.2 Feelings of Inadequacy

Participants described a lack of confidence for both mathematics and science, although most spoke of science as fun experiments and very few spoke of the mathematics taught prior to kindergarten as having the potential to be fun.

> I find that, especially the science, that I actually like it. Because some of the activities, I have fun as well. It's real fun doing it with the kids. I think because I've stayed in the field so long and I've done a lot of training, I have opened myself up to math and science a lot more. (Interview 3, p. 6)

Often, when beginning the interview, participants stated that they did not see themselves as 'a math person' and did not necessarily identify a connection between their own feelings about mathematics and their instructional practice.

> Because on this level, what I'm doing with them, I can do that on that level. Now if you put me in a college level working with older people, let's say college level, any college level, or either in a high school level, I wouldn't be able to do that. But on this level, yeah, I can do this because we're not doing any that kind of math. We're doing just simple math. I can't explain to you exactly how we are incorporating math in the classroom, in the different areas, but I know we're doing math. And I know I can do the math with them. On this level, I can do it. (Interview 2, p. 6)

When asked about their own encounters with mathematics, often participants were negative about their experiences as students, either pinpointing a particular instance where their confidence in their own abilities dropped or highlighting a general lack of positive mathematics experiences throughout their education. 'I had a second grade teacher tell me that I was stupid and that I would never be able to learn math. And I have been afraid of math ever since' (Interview 5, p. 7).

> When I first started school, I loved it. And then I had an incident with my dad, where he was drilling me. And even though I knew the answer, the more he badgered me for the answer, I changed my answer and he was angry. And made me fearful of doing math. It made me afraid. And I always felt that no matter how hard I tried, and knowing the answer, I would always give, I would always feel afraid. I'm going to fail. I'm going to fail. I'm going to fail. (Interview 6, p. 6)

All participants, however, described a higher level of confidence and value for teaching literacy to young children when compared to mathematics and science. 'We do a lot of language and literacy in that classroom, and they see us writing all the time. They see us writing finger plays on big butcher paper and stuff like that. And we're using different markers and pencils and pens. And I think that makes them want to do it' (Interview 3, p. 1). This finding is consistent with literature that points to an emphasis on literacy over mathematics in teacher preparation and professional development (Lee & Ginsburg, 2007; Linder et al., 2016).

9.4.3 Goals for Instruction

When asked how they might achieve their role as a mathematics educator for young children, participants often indicated the need for mathematics tasks to be meaningful, in that they should be grounded in play-based tasks that connect to children's worlds. When asked for examples, however, participants often spoke of inserting counting opportunities during play, during circle time, or while reading a literature selection, all of which are excellent examples but not necessarily connected to children's worlds.

> In cleaning up, getting them, each one of them have so many things that they need to pick up in order to clean up their area. So if there's four people in the house area, each one of them have to pick up at least one or two things to make sure that they're off the floor. (Interview 6, p. 5)

This apparent focus on counting was also present when participants defined what mathematics should look like for young children. 'I think have a solid foundation of numbers from 1 to 10' (Interview 7, p. 2).

In addition, participants stressed that their role involved getting children ready for formal schooling. They did not see their role as promoting utility value for mathematics or for making connections that would lay the foundation for future mathematics success. For example, participants often identified making patterns as an example of the mathematics they might do in child care settings but never described why making patterns is important or how it connects to the content children will experience at the elementary level and beyond. 'Because when they go to school in kindergarten, they do patterns. And they do math. And we try to teach them the stuff that we know they need to know by the time they get in K5' (Interview 1, p. 4). Further, definitions of what math are important for young children to learn centered on content. Only one participant provided examples that centered on mathematical processes such as questioning or problem solving. This view of what it means for a child to be 'ready' for formal schooling, or kindergarten in the USA, was particularly interesting. As there are differing views on school readiness in the research literature (van Oers, 2010; Wilson, 2009) and school readiness is a particular emphasis in Head Start settings (ECLKC, 2016), we are presently examining how Head Start teachers and families view and define school readiness (e.g., Emerson, 2016). In this work, we have found that Head Start teachers highlight the importance of skill development over all other aspects of cognitive development and place a lesser level of importance on development within the social, emotional, physical, or communicative domains (Emerson, 2016).

9.5 Navigating Future Professional Development

Once we had a better understanding of how this group of teachers identified themselves and their roles as early childhood mathematics educators, we examined each teacher individually to determine how their mathematics identity compared to examples of instruction that were gathered throughout the year. We had observed and recorded each participant a minimum of five times, at different points of the day, from October to April. We highlight two of these teachers in this chapter to portray how connecting these ongoing observations with a deeper understanding of how they viewed themselves helps us to better support their practice by making more informed decisions about their future professional development experiences. Prior to the BEEMS professional development, these teachers did not necessarily see math as important for children in child care settings. By the end of the year, these teachers described the importance of providing mathematical experiences for their students, even if they lacked the overall confidence to do so. We ground our examination of these two teachers' identity and enactment of practice in Bloom's taxonomy within the affective domain (Anderson, & Krathwohl, 2001; Krathwohl, 2002). This well-established framework provides a means for us to examine where each teacher falls in terms of their dispositions toward early childhood mathematics and how these dispositions affect their practice.

9.5.1 *Sherry*

Sherry is the lead teacher in a toddler class and center manager for one of the Early Head Start sites. The students in her class begin at approximately 25–27 months and typically turned three by the end of the school year. Sherry has been teaching young children for over 20 years, 11 of those years with Early Head Start, and has a Bachelor's degree in Counseling Education and a Master's degree in Early Childhood Education. In our interview, without being prompted, she confidently described her motivation in working in childcare. 'I wanted to make a difference and Head Start was it. And when I came… it was challenging. But I saw where I made a difference, having the parents recognise that their children were beginning to learn different things. It made a difference in my life. It made me think that I was doing something great.' (Interview 6, p. 1)

When Sherry first entered into the BEEMS professional development group, she was quite wary of whether or not infants and toddlers should be engaging in mathematics experiences. Her primary focus within the classroom was to promote autonomous behavior and social/emotional growth. When we discussed what she learned from the experience, she described how surprised she was at the children's growth throughout the year. In one example, she describes a child identifying one less in a representation of a pizza:

> I can tell you, yesterday in the house area, there's pizza parts over there and there's pretend pizza over there. And one of the girls came over and said there's a piece missing. And I said, there's a slice of pizza missing? And she said a slice of pizza is missing. I said, yes. I said, so how many pieces do you have? And she began to count them. I said so how many are missing? She thought about it. And I said, well count them. And I said how many spaces do you have for slice of pizza? She said four. And I couldn't believe that she remembered that. So I said how many pieces are missing? One. So I'm like oh my goodness. (Interview 6, p. 2)

When describing how she viewed her role as a mathematics educator following the year of professional development, she reflects:

> I've grown a lot because most, I started out directing everything. And now I can step back and allow them to move throughout the environment without me having to interject. They usually self-regulate. And they correct each other. So I don't even have to say a word most days. They're correcting each other. A facilitator. I facilitate more than direct. (Interview 6, p. 3).

The following vignette describes an example of the emphasis on didactic instruction that she was reflecting upon in her interview.

During free play in the toddler classroom, Sherry is working with two eighteen-month-olds to stack cups with a standard number on each face. As they are working to stack the cups, Sherry directs them to stack certain cups in a certain way rather than letting them explore on their own.

> Sherry: 'How about this one? (*Hands a child a cup and has them stack it.*) It keeps falling. Try another one, try another one (*repeats this statement until the child takes the cup from her hand and stacks it on the column of cups*). Put the green one on, put the green one on (*child stacks the green one*). Let's try another one. Try it. It's almost there, you have to turn it just a little bit. Let's turn it this way instead (*supports child to rotate the cup until it is in place*). Try that one. Try that one. What comes next? Six, five, (*calls out the numbers on each cup's face as she hands him the cups to stack*), four, three, two. (*The stack falls and kids laugh.*)' (Observation 2)

In this vignette, it is clear that Sherry is including mathematical talk in her classroom and encouraging her children to think about the stable order of numbers. However, she is giving children explicit instructions on how to stack the cups, which cups to use, and what to do if the cup does not stay in place. The child is not working autonomously (one of her main goals for social/emotional growth, but not necessarily for cognitive growth). In terms of professional development, Sherry's ability to reflect on her use of didactic instruction (as she described in her interview) indicates a change in her mathematics identity. There were few examples during observations, however, of her being able to allow children to take charge of their own learning through play. For this reason, a goal for future professional development would be to address this disconnect between her desired role and her actual role as a mathematics educator. While didactic instruction is not necessarily negative for children, progressive approaches to early childhood mathematics call for teachers to act as a facilitator, encouraging mathematical growth through play (NAEYC, 2002).

As consistent with the other participants, Sherry spent the majority of time during observations focusing on counting concepts. At the end of the year, when asked to describe what math should look like in a toddler setting, she advocated for integration throughout the day.

> In everything that we do really, math is a part of it, day-to-day living. Whether it's brushing your teeth or combing your hair or just cooking, all of that math is included in all of that in some way or another. (Interview 6, p. 5).

The following vignette portrays a common occurrence in Sherry's class, where mathematics content (particularly geared toward counting) was connected to informal learning opportunities such as snack time.

Sherry is sitting down at snack time with four children ages sixteen to nineteen months. One child is hearing impaired, so Sherry is communicating verbally and with sign language.

> Sherry: How many, how many do I have? One, two, three, four, five- I have five cookies. How many do you have Alex? Can you count them?
>
> Alex: One. (*Sherry repeats the number after him every time he says a number*). Two. Three.
>
> Sherry: It's three! I have five, you have three. (*One child points to Sherry's cookies*). You want one more? Can you eat those and I'll give you more. (*Child counts his cookies and says there is two and then takes a bite of one cookie*). Now there is one and a half.

A few minutes pass.

Sherry: How many bowls are on the table? One, two, three, four (*children repeat as she counts and points to each bowl*). There are four bowls on the table.

Diana: Eleven.

Sherry: Eleven! There's only four (*while laughing*)!

It is unclear from ongoing observations if Sherry was focusing solely on counting skills because she believed this to be appropriate for toddlers or if she did not have a clear understanding of the potential mathematical content that can be appropriate for very young children. Future pathways for professional development could push Sherry to think beyond counting to try to implement a wider scope of mathematics content in her toddler classroom.

9.5.2 Joann

Joann is also a center manager and lead teacher for a class of four-year-olds. She has a wealth of experience with 11 years working at Head Start and is working toward her Master's degree in Early Childhood Education. Joann spends much of her time trying to instill a sense of confidence in her students and encouraging the idea that they can be anything they want as long as they work hard. During our first visit to her classroom following day one of professional development, she had a bulletin board with the title 'A Good Mathematician' with pictures of her students engaging in play that promotes mathematical thinking. This bulletin board demonstrated her commitment to emphasizing mathematics but also her commitment to making math meaningful for her students by showing them how what they do now can build the foundation for their future. When reflecting on the outcomes of the year, Joann was also surprised at what her children could do:

> It's just kind of all coming out because at the beginning of the year, and when you guys where here, it was like oh my god, how are we going to teach this? Is it going to sink? But I see the results, especially the ones that came back this year that was a part of that last year. I see the difference in how they have evolved in their learning and how they're problem solving and how they're thinking about things before they come and ask us or get us to assist them with stuff. (Interview 3, p. 3)

During observations, it was clear that Joann was working toward her goal of implementing mathematics throughout the day (as she stated in her interview). In addition to including mathematical talk in formal learning opportunities such as circle time or small group tasks, Joann used counting as a means to transition from one task to another. For example, during our first visit, Joann had each child choose a number from a bucket, say the name of the number, and find a matching number somewhere in the room. During the next visit, she had children sing a song that called for each child to say a number and for the group to count up to that number. When we asked Joann about her feelings about mathematics, she placed an emphasis on processes over content and focused on the idea that problem solving was a priority.

> Math is problem solving, whatever way you look at it. And if they can get that problem solving skill down early, whether through math, it's going to carry out throughout all other areas of their life. Through social-emotional, social development, interacting with friends, how do we solve this problem? How do we get along with others? How can we make it right? Not just figuring out how this, what number and this number make, how we get that answer, but how can we solve problems? Whatever it may be, whatever part of their life, I think math helps in solving problems and it gives them that skill to do that. (Interview 3, p. 5)

However, when examining videos of her practice, Joann spent the majority of time focusing on procedures, specifically ones associated with counting. She asked a wealth of questions (which was not common for most participants), but often these questions were subject centered (with only one correct answer) rather than person or process oriented (with children able to provide a variety of answers). As she progressed throughout the year, however, we did begin to observe her enacting more mathematical processes during play experiences. In the following vignette, Joann encourages a child to describe how she could verify that a set of numbers represented eight. Rather than just telling her to count fingers, Joann attempted to ask questions that would encourage the child to think of a strategy for determining the total set of fingers.

Joann is sitting in the house area with a group of children during center time. A child (Mya) sees Joann hold up eight fingers (five on one hand and three on the other) and begins to hold up eight fingers. She taps Joann on the shoulder for her to see.

Joann: What's that?

Mya: Eight.

Joann: How do you know that is eight?

Mya shrugs.

Joann: How could you find out? How could you find out if you are right? You could what?

Mya: I could go to you?

Joann: No. If I have three fingers and five fingers, how do I know that I have eight fingers?

Mya: Because we put numbers up there?

Joann: Look at my fingers. How do you know that this is eight? How can you find out if it is right?

Mya: Because I saw you do it.

Joann: You saw me do it- but I could be wrong. So how would you know if I was right? What could you do to find out if I was right? What could you do with my fingers? Count them.

Mya counts the fingers Joann is holding up.

Joann: That's how you know if I am right- you would have to count them. I know you know that is eight but if you want to really make sure that you are right you could do what?

Mya: Count them.

This vignette portrays Joann beginning to consider how to promote more opportunities for process-oriented thinking throughout the day. Previously, Joann might have just told Mya to count her fingers and tell her the total amount in the set. In this example, she is moving past assessing counting skills and focusing instead on building Mya's reasoning skills. For future professional development, Joann's pathways should lead toward finding more ways to increase the amount of opportunities for process-oriented thinking and discussions within her classroom setting.

Unlike many other participants, Joann also recognized mathematics content beyond numbers that could be focused on when working with four-year-olds. In particular, Joann had an affinity for the block center and the connections that could be made between mathematics concepts and block play. In the following vignette, Joann encourages a child to make predictions about the number of blocks it will take to make a square-shaped chimney.

Joann moves over to the block center to observe children building and racing with cars. She sits down with two boys, as they play with large cardboard blocks.

Joann: What are you doing over here? Are you building your chimney now? (*Child nods yes.*) Ok build your chimney. Which blocks are you using? Are they the square ones, the rectangle ones, the star ones, what are they?

Ronnie: A square chimney?

Joann: A square chimney?

Ronnie: Yes.

Joann: Ok, so how many blocks do you have to put together to make a square chimney?

Ronnie: Two! (*holds up two fingers*).

Joann: One, two? Just two blocks? Ok, let me see you make it and then we can talk.

Ronnie puts four rectangular prisms together in the shape of a square. His friend John comes to help.

Joann: So how many blocks did you use to make your chimney?

Ronnie: Three (*holds up three fingers*).

Joann asks him to count, he counts each block showing one to one correspondence and changes his response to four.

Ronnie: Now we need to put these blocks on top (*holding up a smaller rectangular prism that is half the size of the blocks he was originally using*).

Joann: Ok - what are we going to do now?

Children begin to explore stacking smaller blocks on top of larger blocks.

While it was exciting to see Joann incorporating more mathematical interaction in the block center, she did display some misconceptions about the names of shapes (i.e., calling a block the 'rectangle one' rather than rectangular prism). Future professional development pathways could determine if this was, in fact, a misconception or an attempt to make math vocabulary more attainable for young children. Further, even though the children began to make connections in regards to the attributes of shapes (by comparing the shape and size of a large and small rectangular prism), Joann focused on the counting connections she could make as children explored. Future professional development could have her reexamine this interaction to determine what other questioning opportunities she could have enacted.

9.6 Making Connections

For transformative professional development to, in fact, be transformative, we as providers must meet the needs of our participants from where they are, rather than from an idealized point of view that is not grounded in context. Understanding how participants view themselves can be a tremendous asset when attempting to make professional development experiences meaningful (i.e., Borko, 2004; Loucks-Horsley et al., 2010). At the beginning of this study, none of the participating teachers described feeling confident in their abilities or necessarily agreed with the importance of promoting mathematics experiences for young children. They saw mathematics as procedurally oriented and drew from their own experiences as students when thinking about how to incorporate more mathematics into their classroom environment. These perceptions could stem from participants seeing themselves as caregivers or as literacy educators and not necessarily seeing themselves as mathematics educators. Hobbs (2013) and Spillane (2000) highlight this notion of teaching 'out-of-field' and how experiences with teaching subject matter that is essentially out of their comfort zone might influence their identity. In this case, while teachers may not describe themselves as early childhood mathematics educators, the idea that mathematics is a part of their role emerged for all participants as they progressed throughout the year. They all tried to implement recommended practices and were surprised at children's ability and growth overtime. 'At the time, you think, at least I thought, is this working because I'm like I don't see it. But now I see it' (Interview 3, p. 3). They became open to the process of changing and working with facilitators of professional development to create reflective plans for professional development. One year later, some now advocate for increased focus on mathematics in their new school settings:

> And that training in particular has stayed with me, and I use a lot of those things in my classroom. I've taken them with me. I know when I came to this Head Start, they were, I have a teacher that's been working, she's 62. My TA is 62 and she kept saying this is not math. She was really upset. She said we don't do it like this in Georgia. You all might do it that way in South Carolina. She had an issue, but she got over it. I kept saying I know it's different, but they need this for school readiness. They need this. And then we had a workshop about three weeks ago that talked about STEM in Georgia and how important STEM is. And everybody looked at me and they went, and they stopped fussing…They [co-workers] didn't like me because they felt like I was trying to do too many outlandish things. They kept saying we never did this like this. The battle was with the adults. The battle is not with the kids. (Interview 5, p. 7)

While all of the participants are on different pathways, these pathways continue to move forward as these educators refine their mathematics identities.

9.7 Conclusion

The overarching theme of this book is making connections, and we address connections and disconnections between early childhood teachers' mathematics identity and their enacted classroom practice. We also address the connections that are implicit between identity, instructional practice, and pathways of professional development. Transformative professional development geared toward building reflective practitioners in the field of early childhood mathematics should capitalize upon an understanding of not only what participants do within the classroom but also who they are as early childhood mathematics educators. These explorations into identity, enactment of practice, and professional development connect explicitly to our past and present work in mathematics education. Whether it be examining the role of the facilitator in motivating teachers to be engaged during mathematics professional development (Linder et al., 2013; Linder, 2011); examining communities built through ongoing mathematics professional development (Linder, 2012); analyzing mathematics professional development opportunities available to early childhood educators (Linder et al., 2016; Simpson & Linder, 2014); or examining how professional development influences classroom practice specific to issues of gender (Simpson & Linder, 2016), student motivation (Linder et al., 2015), or reflective practice (Linder, 2010), identifying and understanding the connections that can exist between the facilitator, teacher, and student are profoundly important for implementing effective professional development experiences.

The next steps for this work are currently underway and involve further investigation into how individual pathways of professional development can continue to influence these participants' mathematics identity and determine the resulting impact on instructional practice and on young children's understanding of and dispositions toward mathematics. This individualized approach to professional development takes into consideration common trajectories that early childhood teachers may progress through in terms of mathematics understanding and connections to practice, just as children progress along trajectories of learning within mathematics content (Clements & Sarama, 2004; Clements et al., 2011). Future research must consider the commonalities that early childhood teachers experience in terms of previous education, beliefs, and contextual issues related to mathematics instruction in addition to the differences that are prevalent when developing a comprehensive transformative professional development framework for early mathematics. This framework should determine overall principles and standards for early childhood mathematics professional development and teacher preparation, which (due to various barriers present in birth to five settings) have the potential to be quite different than professional development in the elementary years and beyond. In particular, research regarding the feasibility of individualized pathways for professional development in birth to five settings is suggested.

References

Anderson, L. W., & Krathwohl, D. (2001). *A taxonomy for learning, teaching, and assessing: A revision of bloom's taxonomy of educational objectives*. New York: Addison Wesley Longman.

Boaler, J. (2002). The development of disciplinary relationships: Knowledge, practice and identity in mathematics classrooms. *For the Learning of Mathematics*, *22*(1), 42–47.

Borko, H. (2004). Professional development and teacher learning: Mapping the terrain. *Educational Researcher*, *33*(8), 3–15.

Butler-Kisber, L. (2010). *Qualitative inquiry: Thematic, narrative and arts-informed perspectives*. Los Angeles, CA: Sage.

Chronaki, A., & Matos, A. (2014). Technology use and mathematics teaching: Teacher change as discursive identity work. *Learning, Media & Technology*, *39*(1), 107–125.

Clements, D. H., & Sarama, J. (2004). Learning trajectories in mathematics education. *Mathematical thinking and learning*, *6*(2), 81–89.

Clements, D. H., Sarama, J., Spitler, M. E., Lange, A. A., & Wolfe, C. B. (2011). Mathematics learned by young children in an intervention based on learning trajectories: A large-scale cluster randomized trial. *Journal for Research in Mathematics Education*, *42*(2), 127–166.

Cobb, P. (2000). Conducting teaching experiments in collaboration with teachers. In A. E. Kelly & R. A. Lesh (Eds.), *Handbook of research design in mathematics and science education* (pp. 307–334). Mahwah, NJ: Law rence Erlbaum Associates.

Drake, C., Spillane, J. P., & Hufferd-Ackles, K. (2001). Storied identities: Teacher learning and subject-matter context. *Journal of Curriculum Studies*, *33*(1), 1–23.

Emerson, A. (2016). Head Start parents' beliefs about school readiness and their implications for practitioners and teacher educators. Presented at the annual meeting of the National Association of Early Childhood Teacher Educators. Baltimore, MD.

Gellert, U. (2008). Routines and collective orientations in mathematics teachers' professional development. *Educational Studies in Mathematics*, *67*, 93–110.

Gujarati, J. (2013). An "inverse" relationship between mathematics identities and classroom practices among early career elementary teachers: The impact of accountability. *The Journal of Mathematical Behavior*, *32*, 633–648.

Gunderson, E. A., Ramirez, G., Levine, S. C., & Beilock, S. L. (2012). The role of parents and teachers in the development of gender-related math attitudes. *Sex Roles*, *66*(3–4), 153–166.

Hobbs, L. (2013). Teaching 'out-of-field' as a boundary-crossing event: Factors shaping teacher identity. *International Journal of Science and Mathematics Education*, *11*, 271–297.

Krathwohl, D.R. (2002). A revision of Bloom's Taxonomy: An overview. *Theory Into Practice, 41*(4).

Lee, J. S., & Ginsburg, H. P. (2007). Preschool teachers' beliefs about appropriate early literacy and mathematics education for low-and middle-socioeconomic status children. *Early Education and Development*, *18*(1), 111–143.

Linder, S. M. (2010). A model of planning for mathematics instruction in the 21st century: Learning from the Reggio Emilia approach. *Teaching Children Mathematics*, *17*(4), 249–254.

Linder, S. M. (2011). The facilitator's role in elementary mathematics professional development. *Mathematics Teacher Education and Development*, *13*(4), 44–66.

Linder, S. M. (2012). Building content and communities: Developing a shared sense of early childhood mathematics pedagogy. *Journal of Early Childhood Teacher Education, 33*(2), 109–126.

Linder, S. M., Eckhoff, A., & Igo, L. B. (2013). Identifying influential facilitators of mathematics professional development: A survey analysis of elementary school teachers. *International Journal of Science and Mathematics Education*, *11*(6), 1415–1435.

Linder, S. M., Rembert, K., Simpson, A., & Ramey, M. D. (2016). A mixed-methods investigation of early childhood professional development for providers and recipients in the United States. *Professional Development in Education*, *42*(1), 123–149.

Linder, S. M., Smart, J. B., & Cribbs, J. (2015). A multi method investigation of mathematics motivation for elementary age students. *School Science and Mathematics, 115*(8), 392–403.

Loucks-Horsley, S., Stiles, K. E., Mundry, S., Love, N., & Hewson, P. W. (2010). *Designing professional development for teachers of science and mathematics* (3rd Ed.). Thousand Oaks, CA: Corwin Press).

Margolinas, C., Coulange, L., & Bessot, A. (2005). What can a teacher learn in the classroom. *Educational Studies in Mathematics, 59*(1), 205–234.

Moerer-Urdahl, T., & Creswell, J. W. (2004). Using transcendental phenomenology to explore the "ripple effect" in a leadership mentoring program. *International Journal of Qualitative Methods, 3*(2), 19–35.

Moustakas, C. (1994). *Phenomenological research methods*. Thousand Oaks, CA: Sage.

Nasir, N. I. S. (2002). Identity, goals, and learning: Mathematics in cultural practice. *Mathematical Thinking and Learning, 4*(2–3), 213–247.

National Association for the Education of Young Children (NAEYC). (2002). *Early childhood mathematics: Promoting good beginnings.* Retrieved from http://www.naeyc.org/files/naeyc/file/positions/psmath.pdf.

National Council of Teachers of Mathematics. (2000). *Principles and standards for school mathematics.* Reston, VA: NCTM.

National Research Council (NRC). (2001). *Adding it up: Helping children learn mathematics.* J. Kilpatrick, J. Swafford, and B. Findell (Eds.). Mathematics Learning Study Committee, Center for Education, Division of Behavioral and Social Sciences and Education. Washington, DC: National Academy Press.

Norton, A. H., & McCloskey, A. (2008). Teaching experiments and professional development. *Journal of Mathematics Teacher Education, 11*(4), 285–305.

Office of the Administration for Children and Families Early Childhood Learning and Knowledge Center (ECLKC). (2016). *History of head start.* Retrieved from https://eclkc.ohs.acf.hhs.gov/hslc/hs/about/history.

Sachs, J. (2005). Teacher education and the development of professional identity. In P. Denicolo & M. Kompf (Eds.), *Connecting policy and practice: Challenges for teaching and learning in schools and universities* (pp. 5–21). New York: Routledge Farmer.

Simpson, A. (2015). *Kaleidoscopic view of voices shaping female and male adolescents' dynamic mathematics identity within single-sex and coeducational environments* (Doctoral dissertation). Retrieved from ProQuest. (UMI Number 3707202).

Simpson, A., & Linder, S. M. (2014). An examination of mathematics professional development opportunities in early childhood settings. *Early Childhood Education Journal, 42*(5), 335–342.

Simpson, A., & Linder, S. M. (2016). The indirect effect of children's gender on early childhood educators' mathematical talk. *Teaching and Teacher Education, 54*, 44–53.

Smith, M. S. (2001). *Practice-based professional development for teachers of mathematics.* Reston, VA: National Council of Teachers of Mathematics.

Spillane, J. P. (2000). A fifth-grade teacher's reconstruction of mathematics and literacy teaching: Exploring interactions among identity, learning, and subject matter. *The Elementary School Journal, 100*(4), 307–330.

Sztajn, P., Hackenberg, A. J., White, D. Y., & Alexsaht-Snider, M. (2007). Mathematics professional development for elementary teachers: Building trust within a school based mathematics education community. *Teaching and Teacher Education, 23*(6), 970–984.

Van Oers, B. (2010). Emergent mathematical thinking in the context of play. *Educational Studies in Mathematics, 74*(1), 23–37.

Wilson, D. M. (2009). Developmentally appropriate practice in the age of testing. *Pediatrics, 123*(2), 431–436.

Authors Biography

Sandra M. Linder is an Associate Professor in Early Childhood Mathematics Education at Clemson University. Her research centers on enhancing teacher quality in mathematics through professional development and promoting young children's understandings of and dispositions toward mathematics. She has worked extensively with in-service and pre-service mathematics educators of young children from birth to age ten. Dr. Linder has multiple national and international publications relating to early childhood and elementary mathematics. She was a coauthor of the Math Out of the Box™ K-5 curriculum and is currently a lead author on the mathematics portion of a STEAM curriculum for preschool children.

Amber M. Simpson received her Ph.D. in Curriculum and Instruction, Mathematics Education from Clemson University in 2015. Amber's research interests include understanding individual's STEM identity, specifically female's identification or dis-identification in STEM-related degrees and careers, and how one's STEM identity may transform due to different formal and informal learning experiences. She is also committed to understanding the ways in which current and prospective mathematics teachers learn to teach as an individual within a social-cultural-historical collective unit. Through peer-reviewed journal publications and presentations at national and international conferences, she is an active member of the mathematics education community.

Chapter 10
Using Mathematics to Forge Connections Between Home and School

Tracey Muir

Abstract This chapter reports on two initiatives that focused on improving parental understanding of contemporary mathematical practices as enacted in their child's school and classroom. The first initiative entitled Engaging Families in Numeracy, commenced in 2009, and involved students from a variety of classes, in two different schools. The project was aimed at engaging parents in numeracy-related activities with their child at home on a regular, ongoing basis. Parents were initially surveyed in order to determine their knowledge and perceptions of how their child was taught numeracy and then invited to provide regular feedback on the activities their child engaged in at home. While parents were generally supportive of the programme, it was felt that there was still a lot of uncertainty about what actually happened in mathematics lessons at school. To address this, a second initiative, the 'Maths Club' was developed, whereby parent workshops were conducted by the author with a focus on familiarising parents with the mathematics curriculum, developing their own mathematical content knowledge, and providing them with opportunities to participate in classroom observation sessions.

Keywords Mathematics · Home school partnership · Numeracy
Parents · Early years of school

10.1 Introduction

> "Even though it looked like you were just playing a game you could see the benefit of the numeracy and the maths skills that were in it"
>
> [Parent comment from Engaging Families in Numeracy project]

There is a growing understanding that when parents, students and teachers work together, there are more opportunities for learning to occur. While most schools and

T. Muir (✉)
College of Arts, Law and Education, University of Tasmania, Launceston 7250, Tasmania, Australia
e-mail: tracey.muir@utas.edu.au

V. Kinnear et al. (eds.), *Forging Connections in Early Mathematics Teaching and Learning*, Early Mathematics Learning and Development,
https://doi.org/10.1007/978-981-10-7153-9_10

communities recognise the importance of establishing collaborative partnerships, it can sometimes be challenging to facilitate practices that enable these partnerships to occur. This chapter reports on two initiatives that focused on improving parental understanding of contemporary mathematical practices as enacted in their child's school and classroom. The first initiative entitled Engaging Families in Numeracy, commenced in 2009, and involved students from a variety of classes, in two different schools. The project was aimed at engaging parents in numeracy-related activities with their child at home on a regular, ongoing basis. Parents were initially surveyed in order to determine their knowledge and perceptions of how their child was taught numeracy and then invited to provide regular feedback on the activities their child engaged in at home. While parents were generally supportive of the programme, it was felt that there was still a lot of uncertainty about what actually happened in mathematics lessons at school. To address this, a second initiative, the 'Maths Club' was developed, whereby parent workshops were conducted by the author with a focus on familiarising parents with the mathematics curriculum, developing their own mathematical content knowledge, and providing them with opportunities to participate in classroom observation sessions.

This chapter begins by providing an overview of the research literature in relation to parental involvement and parental knowledge of contemporary classroom practices. Details of the questionnaire that was used to gauge these aspects in relation to the parents who participated in the project are then provided, along with the findings from their responses. Each of the initiatives is then described, before presenting the conclusions and implications.

10.2 Background

10.2.1 Parental Involvement

In a comprehensive study on family school partnerships undertaken in 2006, Muller (2006) identified seven key principles which directly influenced the success of family school partnerships. Two of the principles which are particularly relevant for this study were the influence of connecting home and school learning and communication. Research has shown that parental involvement affects student achievement (Sheldon & Epstein, 2005) and that students' learning is maximised when strong educational partnerships between school, community and home are developed (Groves, Mousley, & Forgasz, 2006; Vincent, Stephens, & Steinle, 2005). This is consistent with other research which shows that students performed better academically and had more positive school attitudes if they had parents who were aware, knowledgeable and involved (Epstein, 1992, as cited in Anthony & Walshaw, 2007). While many schools encourage parental involvement through open access to classrooms or participation in school events and parent evenings, active parental involvement with mathematics programmes is less prevalent.

Perhaps one of the reasons for lack of parental involvement is related to the subject of mathematics itself. Many adults, in relation to mathematical tasks, admit to feelings of anxiety, helplessness, fear and dislike (Haylock, 2007) and parental comments such as 'I was never any good at mathematics' help to reinforce this view for the next generation. It is an unfortunate reality that many adults and parents today attribute their lack of confidence in mathematics to negative experiences they had throughout their schooling. Through informing and encouraging parents to become more involved in their children's mathematical learning, it may even be possible for parents to act as advocates in promoting reform practices, with an emphasis on process, rather than product, and conceptual, rather than procedural, understanding. Parents can and should have a role to play in reforming mathematics (Peressini, 1998), as they have their own expertise and unique knowledge about their children and thus can contribute to their children's mathematical development. This view is particularly endorsed by Civil and colleagues (e.g., Civil, 1998), who have attempted to address lack of parental involvement through recognising that parents are valuable resources who can be utilised for mathematical instruction. Her research has focused on linking home and school, particularly in the context of working class and immigrant families. Capitalising on their 'funds of knowledge' (Moll, Amanti, Neff, & Gonzalez, 1992), Civil's work valued the rich cultural and cognitive resources that children brought to the school settings. Through engaging parents in workshops and information sessions, she was able to familiarise parents with contemporary practices through modelling reform-based activities, encouraging use of non-traditional approaches with a focus on investigation—a contrast to what many of them experienced in their own schooling.

While it is a common practice to participate in 'home reading' activities, there are less opportunities to engage in organised at home mathematics activities, other than the traditional mathematics homework that occur in primary and secondary school. In terms of providing specific opportunities for parents to be involved in their children's mathematics learning, Sheldon and Epstein (2005) found that a number of involvement activities were effective, including evening workshops involving both parents and children, and provision of teacher-designed interactive homework and mathematics materials for families and students to use at home. Other authors have reported on the use of 'take home packs' (e.g., Goos & Jolly, 2004), maths back packs (Orman, 2000) and 'numeracy bags; (Muir, 2009). An evaluation of 14 experimental family numeracy programmes in Britain noted that the programmes found to be most effective had three key strands: joint and separate sessions for parents and children, a structured numeracy curriculum and bridging activities that parents could use to develop their child's numeracy at home (Anthony & Walshaw, 2007). The findings from these studies show that the practices were an effective means of communicating about mathematics between home and school, contributed to increased participation in mathematically related activities at home and helped to build an increased understanding of contemporary mathematics classroom practices.

10.2.2 Parental Knowledge of Contemporary Mathematical Practices

Many parents feel unable to assist with their child's mathematical development as they are not familiar with the mathematics content that their children encounter in mathematics classes (Peressini, 1998). Pritchard (2004), for example, found that many parents felt uninformed about the mathematics curriculum and the teaching methods used in their child's school. There is evidence to suggest, however, that parents are keen to encourage and support their children in their mathematics education, including those from low SES and culturally different backgrounds (e.g., Anthony & Walshaw, 2007; Muir, 2009). According to Eastaway and Askew (2010), there are four main concerns that parents have about contemporary mathematics practices which they phrased as key questions: Why do they do it differently these days? How can I overcome my own fear of mathematics? How can I get my child to enjoy mathematics and be better at it than I was? Why do they (or I) need to know this? While some of these questions become more pertinent as children move through school to older grades, it is important to acknowledge that parents will bring their own set of beliefs and attitudes to any mathematical experiences that they engage in with their children. As these beliefs and attitudes are likely to influence their interactions with their children, it is important to address these beliefs and help parents to become aware of their impact.

Mathematics classroom observations can provide a means for exposing parents to contemporary mathematics teaching. Civil, Quintos, and Bernier (2003) used classroom observations to engage parents in a dialogue on issues around reform-oriented mathematics teaching. They found that there were at least three factors which coloured the parents' experiences when they observed mathematics classes: their involvement in the Math and Parent Partnerships in the Southwest (MAPPS) programme (a programme conducted by the researchers which aimed to illustrate reform-based mathematics), their prior experiences with learning school mathematics themselves, and their experiences with their own children's mathematics instruction. Their observations focused on 'then and now', in terms of comparing what they saw in the classroom to their own experience when they were in school, and the role of 'understanding' in mathematics. Their findings suggested that the parents compared and contrasted their own experiences, and filtered what they observed through their own beliefs and values, in order to appreciate reform practices.

The projects discussed in this paper build on previous work undertaken by the author. The numeracy bag project was reported in Muir (2009, 2012b, 2016), while the maths club was described in Muir (2012a). In addition, a third initiative, 'Counting on the Count' was also detailed in Muir (2016). Common to all these reports was the emphasis on linking school activities with home, together with the expectation and belief that parents would engage in the experiences with their children. This chapter synthesises the findings from previously published reports through expanding upon participants' experiences and highlights the positive impact that the projects have had on the school communities involved. Future research in this area will investigate other ways of building on parental funds of knowledge and evaluating long-term effects of numeracy-related initiatives.

10.3 Eliciting Parental Perspectives

Returning to the initiatives discussed in this chapter, in order to ascertain parents' current understandings of contemporary practices and to investigate their own beliefs and attitudes, a questionnaire requiring Likert responses to 11 items (see Table 10.1) was distributed to parents, prior to participating in the 'take home numeracy bags' programme. As reported in Muir (2012b), the results of the questionnaire were consistent with the findings in the literature in that parents were largely uninformed about contemporary mathematical practices and generally held a traditional view of mathematics. Table 10.1 (reproduced from Muir, 2012b), shows parental responses to the items in the questionnaire. A total of 34

Table 10.1 Parent responses to the items in the questionnaire

Statement	SA/A responses (%) $n = 34$	SD/D responses (%) $n = 34$	*N* responses (%) $n = 34$
I am satisfied with the way I was taught mathematics in school	47	32	21
Maths is about learning the correct procedures to solve problems	91	3	6
There is a 'correct' way to do any maths problem	47	29	24
Mental computation means knowing your tables	47	15	38
I have a good understanding of how my child is taught numeracy in school	32	32	36
I think the way maths is taught in classrooms today is effective	32	3	65
I know what types of mathematical skills and understandings my child has	68	6	26
Games and activities are a good way to learn mathematics	97	0	3
Worksheets and textbooks are a good way to learn mathematics	85	9	6
Telling children the answer is a good way of helping develop their mathematical understanding	15	41	44
I regularly engage in numeracy-related activities with my child	79	18	3

questionnaires were returned from three different early childhood classes; two from School A and one from School B. The responses from both schools were combined. Strongly Agreed and Agreed (SA/A) responses were combined, as were Strongly Disagreed and Disagree (SD/D) responses.

Table 10.1 shows that 91% of parents agreed that 'maths is about learning the correct procedures to solve problems' and that 85% believed that 'worksheets and textbooks are a good way to learn mathematics'. This is consistent with a traditional view of mathematics, and not necessarily reflective of their child's experiences as most of the parents had young children who would not typically have experienced those practices in their classroom yet. Instead, their beliefs are most likely to be a reflection on their own mathematical experiences, or older children's experiences, or a reflection of societal norms. It was encouraging to see that 97% of parents agreed that 'games and activities are a good way of learning mathematics' and that 79% of parents agreed that they 'regularly engage in numeracy-related activities with my child'. In terms of gauging parents' familiarity with contemporary practices, only 32% agreed that they 'have a good understanding of how my child is taught numeracy in school' and 'I think the way maths is taught in classrooms today is effective'. The high number of neutral responses to these items (36 and 65%, respectively) indicates that parents lacked sufficient information to either agree or disagree with these statements. In summary, parental responses to the items indicated that they held mixed views about mathematical practices—on the one hand, they advocated the use of games and activities, yet also believed that maths was about learning the correct procedures to solve problems. Their responses indicated that they were not well informed about mathematical practices. In order to address this, opportunities for parents to engage in authentic mathematical experiences, and exposure to contemporary mathematical practices, were key features of the initiatives discussed in this chapter.

10.4 Take Home Numeracy Bags

As reported in Muir (2009), this initiative was instigated with a grade 1/2 class (aged between 6 and 7 years) of 28 students in a local district high school (Foundation—Year 10) set in a low socio-economic area. The aims of the study were to investigate parents' perceptions of mathematics and current teaching practices. An intervention programme was implemented whereby students took home weekly numeracy packs and engaged in mathematical activities at home. The study was later extended to other classes in the school and to a different District High School (Foundation to Year 10) in 2010.

The intervention programme involved each child taking home a different numeracy activity each week. The activities were designed to be interactive and support the mathematical experiences undertaken in the classroom. Every Monday each child would receive their 'numeracy bag' that would contain activity instructions, materials and guidelines for parents. There was also a short explicit rationale that explained the purpose behind the activity. Figure 10.1 shows an example of one of the activities and how it was presented to parents. The expectation was that the child would engage in the activity 2–3 times over the week with their parents (see Fig. 10.2) and/or other family members, return the activity on Friday and receive a new activity the following Monday. Each activity bag contained a feedback sheet which required parents to provide data about the child's level of engagement with the activity and the mathematical understandings that were revealed (see Appendix).

Fig. 10.1 Example activity sheet

1-100 Chart

You will need:
Cut out pieces of 1-100 chart

What to do:
Place the pieces together to form a chart showing the numbers from 1-100

Purpose of activity:
This activity is designed to develop an understanding of many concepts. By placing the pieces together, children can form an understanding of the place value system and the sequence of two-digit numbers. Once the chart has been assembled, children can be asked to locate patterns in the chart (e.g., look at the multiples of 11—cover with counters, what do you notice?).

Counting with the chart:
The chart can be used to add and subtract two-digit numbers. For example, place your finger on 23. To add 34, move your finger down 3 rows (as each row represents 10) and move your finger across 4 (as each column represents 1). Try adding a variety of numbers in this way. Subtracting two-digit numbers can also be attempted using the chart—e.g., 67—32 can be found by moving up three rows, and across 2 spaces.

36 + 43 =

Fig. 10.2 'Susan' and parent with her take home numeracy bag

10.4.1 Parental Feedback on Weekly Numeracy Activities

The return rate of the feedback sheets was very high; 24 parents of the 28 students in the class regularly completed and returned weekly feedback sheets for their child. A total of 144 forms were returned over the course of the six-week intervention. The comments recorded by parents revealed that they were able to identify and describe some of the mathematical behaviours they observed. For example, with reference to an activity where children had to form pairs of cards that equalled ten, one parent wrote:

> He understood that that he had to add up; counted on fingers at beginning [but] remembered the pairs towards the end without adding

Another activity required children to place counters in designated ways on a ten-frame. The following comments are indicative of the feedback received:

> She placed the counters in a 'logical' way and could easily tell me how many counters she needed to make 10.

> Filling the frame from left to right; instantly recognising how many are on the frame and able to say straight away how many more would take 10

In order to establish the correct sequencing of counting numbers 1–10, some activities focused on arranging cards or numbers in order. Again, parents were able to comment on the mathematical behaviours they observed as the following illustrates:

> He could count to ten easily … he could work out the hidden number by looking at the others not hidden

There was also the provision for parents to indicate how often the activity was undertaken, whether or not it was appropriate for their child's level of understanding, whether or not the child enjoyed the activity and were the instructions clear. Table 10.2 provides a summary of parental feedback received in relation to these aspects. The number of times the activity was undertaken ranged from 1 to 12, with the mean being 3.8 times.

The number of returns (144) itself was promising as it demonstrated that parents were willing to provide feedback and obviously engaged with their child in the weekly activities as they were able to comment on what they found. It was particularly pleasing to see that 98% of parents identified that the instructions were clear, highlighting the importance of providing activities that are easy to carry out. This may also have attributed to the high rate of return and willingness to complete the activities. One of the most challenging aspects was to provide activities that were age appropriate. While 77% of parents indicated that the activities were age appropriate, it was reasonable to expect that some would be deemed too easy or difficult for their children. Statements given to support the selection of 'disagree' on the feedback sheet in relation to 'activity appropriate' included the following:

> He could count to ten easily ['What's missing?' activity]

> It was too hard without help; he couldn't really understand subtraction [Tower blocks activity]

Interestingly, disagreement comments in relation to this aspect did not always result in a disagreement statement for 'enjoyed activity', indicating that enjoyment does not always depend upon success.

Following six weeks of the implementation of the programme, parents were invited to participate in a follow-up interview about their experiences with working with the numeracy bags. Feedback received from the interview participants indicated that they viewed the project positively and would like to see it continued. All participants indicated that it gave them a better understanding of classroom practices, with one comment being, '…even though it looked like you were just playing

Table 10.2 Weekly parental feedback received after completion of activities

	Agreement (%)
Activity appropriate	77
Enjoyed activity	88
Clear instructions	98

a game you could see the benefit of the numeracy and the maths skills that were in it', and:

> Like I've never seen it before … it was quite interesting to know that that's what he does do in the day – that's his maths at school – and to know what he does and where he's at and plus to help him with it – I thought it was very interesting.

Other feedback included the following:

> Yes, it gave my son a better understanding of how to work out the answers to maths problems. I believe that the problems faced by doing these maths questions gave me a better understanding of where my son's level is.
>
> Yes, it showed me how she can think through problems and how playing a game can help with maths.
>
> It was really good for Nigel because he has auditory processing problems and needs to do things over and over again and these activities let him do that

Suggestions for improvement included trying to 'cater for each student's level' as some parents felt that at times the activities were too easy or challenging for their children. Selection of activities that provide access for a range of abilities, together with suggestions for adaptations and extensions, could help to address this issue.

Feedback from the teachers was also positive, particularly in terms of providing a link between school and home:

> Sometimes they'll [parents] come in and say that was a bit hard; but generally really positive comments – saying how we did that or talk about the activity; it's been really good for parents to see the purpose behind what we're doing; it has certainly been worth doing and I'd like to keep it going – maybe in more of a self-serve way – have a bank of about 40 in room and still monitor it to make sure everything is returned – like they do with their home readers.

In summary, feedback from the parents and teachers was very positive, and the students also indicated that they enjoyed taking the activities home each week. The programme achieved its aims in that through regularly engaging in mathematical activities at home, parents could not only see what their mathematical understandings their child had, but also gained a better understanding of what constitutes mathematics in today's classrooms.

10.5 The Maths Club and Classroom Observations

As reported in Muir (2012a), the Maths Club was introduced to the second school, Mountain Ville District High School, in 2010. The club was established to support parents who did not feel confident with helping their children with mathematics at home. It was modelled on the Mathematics Awareness Workshops that were conducted as part of the MAPPS project (Civil et al., 2003). It was open to all parents, however, typically participant were parents whose children were 4–12 years old. Maths Club workshops were organised around areas of 'need', including algorithms, tables and mental computation and fractions. Each workshop followed a similar format, which included provision of some information, hands on activities and games and provision of resources, including follow-up reading and Websites. The workshops were of an informal nature and interaction was encouraged. As most of the sessions were generally targeted towards upper primary mathematical concepts, they will not be described in this chapter. One of the requests from parents, however, was to gain a better insight into what happens in early childhood, primary and secondary mathematics classrooms. With the cooperation of the teachers and classes in the school, the author organised for this to occur. Full consent was given by all participants and ethical considerations such as maintaining confidentiality when observing and participating in focus group interviews were conveyed to all participants. The results of the observation of a Preparatory/Year 1 classroom are described in the next section.

10.5.1 The Preparatory/Year 1 Lesson

10.5.1.1 Lesson Overview

The lesson observed took place in October 2012, with a Preparatory/Year 1 class of 25 students. Two of the five parents observing had children in the class. At the start of the lesson, the students were seated in front of the teacher, Miss Long, who was seated next to a whiteboard. The students were given a task, written on a piece of paper, in which they were to make up story problems involving the subtraction of chickens where the answer would be four. The task was open-ended in that they could use any combinations of numbers and little teacher direction was given. The students then worked on the task for about 20 min, before regrouping on the mat in front of the teacher. Students worked individually on the task but regularly talked with each other throughout the lesson. They recorded their solutions in their maths books and some students shared these towards the end of the lesson when they regrouped. The lesson concluded with a whole class game in which students were encouraged to subitize to add cards with dots on them.

10.5.1.2 Parental Observations

Immediately following the lesson, the parents participated in a focus group interview where they were asked questions about what they observed. The researcher was particularly interested in investigating what aspects of the lesson they noticed, and what aspects of students' mathematical understandings were identified. Parents were ambivalent about the use of an open-ended task to engage the students as the following comments show:

> I thought that by giving them such an open-ended question, to some of those kids, it just went 'whoosh' straight over their heads without giving them, say, 'You've got 10 chickens and you got four left, what happened to the remainder of them?' might have got some of those other ones, but the majority sort of sat there and they needed somebody to go and give them a figure to start with.
>
> [John]

> I think she was trying to get them to think within their ability so children who couldn't think beyond 1 or 2 or more than 4 were going to give answers that were only one or two more than four but the kids who could think to bigger numbers and who knew how to take four away from larger numbers, were going to give bigger answers, so in a way they were self-regulating their learning – it also gave her also an understanding of what their ranges are.
>
> [Bronwyn]

When asked to describe observations of students engaging in mathematics, the parents were in agreement that the drawing of the chickens tended to detract from the focus on subtraction and mathematics in general, with one child not even attempting the task because he could not draw chickens:

> Well there was one boy who didn't – he just drew a great big circle with a big line and a scribble through it and I went over and he had done the numbers, he had done the maths but he hadn't done a picture, and he was just getting frustrated and the lines were getting darker [emphasised] on the page, and I said, why haven't you drawn any chickens? And he said, I can't draw chickens. And I said why don't you get out the counters that you were going to use to count with, and trace them? So he did – he got five – he picked all the same size, he got them out and he traced around one and that got him going and he drew the others and he put beaks on them and he had done the maths to count the counters out to start with and you know – perhaps it was just the one on one – you know – someone taking an interest probably – I don't know – it was interesting that he didn't progress until he decided that he could draw chickens after all.
>
> [Bronwyn]

Bronwyn's comment also shows that she was a participant-observer in the class. All parents talked individually with the students and some assisted them with their work. In the focus interview, some of them raised concerns that some students would not have engaged with the task at all, without some adult intervention, as the following comment illustrates:

> There was one kid there and he had like when they sat down to do the sum, there were some that did it just straight away and there were some that had no idea; one boy there – and I went to him a couple of times – he had done nothing – he was sitting there blank and he just didn't know what to do – then I'd go up and mention a few things and get him started and once he got talking to me, he did it but to start with, he just had no idea
>
> [Julie]

In terms of identifying effective teaching practices, the parents tended to focus on the classroom environment and personal qualities of the teacher. One parent, Ellen, commented that 'she gave the children a lot of opportunities to interact and talk .... and also made it relevant by using chickens'.

When asked whether or not the parents had observed instances of students interacting with each other about mathematics, there was general agreement that this did not occur. The interactions were mainly social, not mathematically focused, as the following comments show:

> I didn't find they interacted between themselves about maths – I think that's probably something that children at that age don't really do – they might share pictures but they tend to work individually even if they're in a group – I'm doing this...
>
> [Bronwyn]

> If you ask them what they've got, how many chickens – they'll tell you but they're not likely to ask their friend – I didn't see any of that – they were focused on their drawing – their chickens, their beaks, their worms; they were focusing on their work, I didn't see them interacting about maths
>
> [Julie]

In the focus interview, the parents were also specifically asked to comment on the practice of sharing at the end of the lesson. It was reasonable to expect that this practice would not have been a feature of their own mathematics lessons, so the researcher was interested in gaining their perspectives on this as a contemporary mathematics practice. The following comments showed that they generally could see the value in this:

> I think it was valuable how she asked them to explain how you got to that answer and the others could listen and see oh there are other ways of doing that – and she got that one girl out to say how she had done it and this one in particular, she used her as an example, then the other kids could see oh yes that's a way of doing it.
>
> [Nina]

> I think it also showed the other kids that at least other people were doing work as well – it wasn't just me – making it part of the collective, that other people did draw chickens and did come up with sums as well – I don't know – it helps them to see where they are with the others and um, yeah, if you're wanting to be as good as somebody – and I don't know if they're at that age – then they might be noticing the quality of other people's work.
>
> [Bronwyn]

Another parent, however, believed that while the students were keen to share their own answers, they were not necessarily interested in listening to others' explanations:

> There's a lot that don't listen, there's a lot that stick their hand up I noticed and didn't get picked – but the majority of kids want to share but there's never enough time to share all of them – they never want to listen to somebody else, they just want to tell you and think once I've shared mine they're not interested.
>
> [John]

Similar in nature to the observations made by the parents in Civil et al.'s (2003) study, parental observations were influenced by their prior school experiences and the experiences they had with their own children's mathematical instruction. One parent, for example, drew upon her knowledge of other practices she had observed by commenting:

> I was surprised when the question was just words – just a strip – because I've seen other worksheets that they've done which have been diagrams or numbers or a combination of both, and I was interested that this was minimal – there was minimal direction given.
>
> [Bronwyn]

Another parent identified that her child would not have participated unless he was prompted:

> Yeah, our boy would have been one of those that would have sat there and gone I don't know where to start.
>
> [Erin]

In summary, the main aspects noticed by the parents included the issue of task differentiation and support for individuals undertaking the task. They tended to focus mostly on the task, but when prompted, also commented on the interactions between students. Further probing could have elicited more information about comparison with their schooling experiences. Interestingly, this aspect was more apparent when the parents observed the upper primary and secondary classroom lessons.

10.6 Conclusions and Implications

The projects documented here helped to address concerns raised in the literature and the survey that parents were largely uninformed about contemporary mathematical practices and felt that they lacked the necessary mathematical knowledge to support their child's numeracy development. The strategies employed in the projects were appropriate for addressing and raising awareness of parents' beliefs and values regarding mathematics teaching and learning and for acknowledging that based on the premise that parents had their own beliefs about mathematics teaching and learning and that parents wanted to support and encourage their children's numeracy. Despite limitations in terms of sample size, the projects illustrate how schools and teachers can establish effective working partnerships with parents to enable them to gain insight into contemporary mathematics practices, which can ultimately improve outcomes for their children. Similar to findings reported elsewhere (e.g., Civil et al., 2003; Muir, 2009, 2012a), parents were appreciative of the opportunity to engage in numeracy-related activities with their children and to be involved in discussions about mathematics teaching.

Part of the success of the projects was the provision of the avenue for communication between both parties. In the numeracy bags at home programme, the purpose of the activities was clearly stated and accompanied by an explanation of the mathematics involved in the tasks. The expectation to complete the feedback sheets provided a mechanism for communication about the activities to occur between parent and teacher and served to make the programme valuable for all parties. In the observation of mathematics classes, parents were engaged in post-lesson observation dialogue where they could discuss what they had observed and clarify their understandings of contemporary mathematics practices.

The numeracy bags at home programme showed that, as Sheldon and Epstein (2005) found, parents were willing to participate in their child's mathematical education and furthermore were able to contribute in a positive way, providing additional insights into their child's development that could be capitalised upon by their teacher. As reported in Muir (2009), while no claims can be made that the six-week programme significantly altered parental perceptions or beliefs about mathematics, it did seem to provide them with an increased understanding of the types of mathematical activities undertaken in the classroom and the mathematical understandings and skills involved with being numerate.

The classroom observations provided another avenue for developing an increased understanding of contemporary mathematics practices. While this practice may be limited in terms of parental participation due to other commitments, it does provide a model of a possible approach to parental involvement (Civil et al., 2003). In terms of implications for further research, ethical considerations need to be made explicit about what is observed in the classroom and shared with the wider community for the parents who take part. In addition, the use of a proforma with targeted areas for observation would assist in guiding the post-lesson discussion.

According to Civil (2001), as parents become more informed about mathematics education issues and more exposed to sound approaches to mathematics teaching and learning, then they may become more active advocates for quality mathematics education. It is hoped that the projects discussed in this chapter have provided an insight into how to involve parents in purposeful experiences and exchanges about contemporary mathematical practices which will help to forge connections between home and school.

Appendix: Numeracy at Home Feedback Sheet

Name of activity:

Child's name:

Date:

How many times did you and/or child complete the activity?

……………………..

This activity was appropriate for my child	Agree	Disagree
My child enjoyed completing the activity	Agree	Disagree
The instructions were clear	Agree	Disagree

What mathematical understandings (or misunderstandings) did your child reveal when participating in the activity?

Do you have any other comments or questions about the activity?

Thank you for taking the time to complete this feedback.

References

Anthony, G., & Walshaw, M. (2007). *Effective pedagogy in mathematics/Pangarau*. Wellington, NZ: Ministry of Education.

Civil, M. (1998). *Parents as resources for mathematical instruction*. Adults learning maths—a research forum, Utrecht, The Netherlands.

Civil, M. (2001). *Redefining parental involvement: Parents as learners of mathematics*. Paper presented at the NCTM research pre-session, April 2001, Orlando, FL.

Civil, M., Quintos, B., & Bernier, E. (2003, April). *Parents as observers in the mathematics classroom: Establishing a dialogue between school and community.* Paper presented at the NCTM research pre-session, San Antonio, TX.

Eastaway, R., & Askew, M. (2010). *Maths for mums and dads.* Great Britain: Square Peg.

Goos, M., & Jolly, L. (2004). Building partnerships with families and communities to support children's numeracy learning. In I. Putt, R. Faragher, & M. McLean (Eds.), *Mathematics education for the third millenium: Towards 2010* (Proceedings of the 27th annual conference of the Mathematics Education Research Group of Australasia, Townsville, pp. 279–286). Sydney: MERGA.

Groves, S., Mousley, J., & Forgasz, H. (2006). *Primary Numeracy: A mapping, review and analysis of Australian research in numeracy learning at the primary school level.* ACT: Canberra.

Haylock, D. (2007). *Mathematics explained for primary teachers* (3rd ed.). Thousand Oaks, CA: Sage Publications.

Moll, C., Amanti, C., Neff, D., & Gonzalez, N. (1992). Funds of knowledge for teaching: Using a qualitative approach to connect homes and classrooms. *Theory Into Practice, 31*(2), 132–141.

Muir, T. (2009). At home with numeracy: Empowering parents to be active participants in their child's numeracy development. In R. Hunter, B. Bicknell, & T. Burgess (Eds.), *Crossing divides* (Proceedings of the 32nd annual conference of the Mathematics Education Research Group of Australasia, pp. 395–402). Wellington, NZ: MERGA.

Muir, T. (2012a). Numeracy at home: Involving parents in mathematics education. *International Journal for Mathematics Teaching and Learning.* Available: http://www.cimt.plymouth.ac.uk/journal/muir.pdf.

Muir, T. (2012b). It's in the bag: Parental involvement in a numeracy at-home program. *Australasian Journal of Early Childhood, 37*(2), 106–112.

Muir, T. (2016). Out of the classroom, into the home. *Teaching Children Mathematics, 22*(8), 497–504.

Muller, D. (2006). *Family-school partnerships project.* Canberra, ACT. Commonwealth of Australia.

Orman, S. A. (2000). Mathematics backpacks: Making the home-school connection. In D. Edge (Ed.), *Involving families in school mathematics* (pp. 51–53). Reston, VA: NCTM.

Peressini, D. D. (1998). The portrayal of parents in the school mathematics reform literature: Locating the context for parental involvement. *Journal for Research in Mathematics Education, 29*(5), 555–583.

Pritchard, R. (2004). Investigating parental attitudes and beliefs in mathematics education. In I. Putt, R. Faragher, & M. McLean (Eds.), *Mathematics education for the third millennium: Towards 2010* (Proceedings of the 27th annual conference of the Mathematics Education Research Group of Australasia, Townsville, pp. 478–485). Sydney: MERGA.

Sheldon, S. B., & Epstein, J. L. (2005). Involvement counts: Family and community partnerships and mathematics achievement. *The Journal of Educational Research, 98*(4), 196–206.

Vincent, J., Stephens, M., & Steinle, V. (2005). *Numeracy research and development initiative 2001-2004.* Canberra, ACT: Commonwealth of Australia.

Author Biography

Dr. Tracey Muir is a Senior Lecturer in Mathematics Education at the University of Tasmania. Her research interests include effective teaching for numeracy, problem solving in mathematics, the use of ICT in mathematics and parental involvement in mathematics education. She has published over 40 journal articles and conference papers and was a previous co-editor of Australian Primary Mathematics Classroom. She is a past Vice President (Development) of Mathematics Education Research Group of Australasia (MERGA). She has presented seminars and workshops at state, national and international conferences and is particularly passionate about working with teachers to enhance their classroom numeracy practices and engaging children in mathematics.

Chapter 11
Young Children's Reasoning Through Data Exploration

Gabrielle Oslington, Joanne T. Mulligan and Penny Van Bergen

Abstract This chapter follows the progress of nine high-ability Year 1 Australian students as they develop reasoning skills through data exploration and analysis. The students used self-portraits drawn by child artists in Kindergarten and Year 3 to develop a rule-based classification model. Students tested their model on larger sets of self-portraits and developed their own illustrations to support the rule-based model. Seven of the nine students demonstrated advanced mathematical reasoning through their analysis of the test findings to inform their judgements regarding the strengths and weaknesses of the model. Students demonstrated this reasoning through graphical representations, reflective statements and two-way tables. These findings demonstrate the potential of rule-based model building and data analysis to extend the mathematical experiences of mathematically gifted young children.

Keywords Gifted education · Statistical reasoning · Model building
Data modelling · Representations · Classification · Early years of school

11.1 Introduction

Reasoning capacities cut across curriculum areas and are essential for functioning in current information-rich and data-driven learning environments. Interpreting data within a specific context and developing inferential thinking are core capacities promoting conceptual understanding, for example across the languages, the arts, the

---

G. Oslington (✉) · J. T. Mulligan · P. Van Bergen
Macquarie University, Sydney, Australia
e-mail: gabrielle.oslington@mq.edu.au

J. T. Mulligan
e-mail: joanne.mulligan@mq.edu.au

P. Van Bergen
e-mail: penny.vanbergen@mq.edu.au

V. Kinnear et al. (eds.), *Forging Connections in Early Mathematics Teaching and Learning*, Early Mathematics Learning and Development,
https://doi.org/10.1007/978-981-10-7153-9_11

humanities, the sciences and mathematics. Young students often start school already demonstrating a wealth of mathematical knowledge and skills (Clements & Sarama, 2009). Recent research indicates that by the first year or two of school, some students have impressive reasoning and analytical capacities. These include the ability to represent categorical and continuous data (Mulligan, 2015) and to demonstrate statistical understanding (English, 2013). In addition, some young students can use contextual information, for example using the details provided in a storybook to predict the likely values for similar items in the story (Kinnear, 2013). Some young students have also been shown to use informal statistical reasoning to form generalizations, using data as evidence (Makar & Rubin, 2009).

In recent decades, educational and developmental research has widely dispensed with the traditional Piagetian position (Inhelder & Piaget, 1958) of young children being incapable of abstract or casual reasoning until they reach adolescence. As early as 1998, for example, Diezmann and Watters demonstrated that rather than being principally temporal, illogical, irrational and pre-causal, children even as young as five could use logical reasoning, rebuttals and exercise deductive reasoning to overcome a previous belief. Drawing on this evidence, Gopnik (2012) proposes that young children in fact reason very like scientists by constantly testing hypotheses against data, by learning from statistics and (informal) experimentation and by observing the world around them. Younger children may in fact find developing hypothesis easier than slightly older children, who may have already developed rigid belief structures which limit the degree to which they are willing to consider alternatives. While it is likely that Piaget underestimated the capacities of most children, young gifted children in particular have been found to show particularly advanced capacity in deductive reasoning and logical inferences (Morsanyi, Divine, Nobes, & Szucs, 2013).

The chapter reports the development of reasoning in high-ability Year 1 students through the process of mathematical model building and data interpretation. It links with the growing body of work on mathematical modelling demonstrating children's capacity to use self-collected data to inform deductive and inferential reasoning.

11.2 Theoretical Background

11.2.1 The Modelling Process

Data modelling, in the context of the mathematics classroom, is a problem-solving process where a system of interest is represented by a mathematical system, which simplifies some aspects and emphasizes others (Zawojewski, 2010). With children enmeshed in a digital and data-driven world, the development of categorization and statistical reasoning through the early years' curriculum is a necessity (English, 2013). The early development of statistical reasoning is generally limited to classroom programmes that give children opportunities to sort and classify materials, where these activities typically involve items with defined attributes such as shape and colour. Irregular and complex objects encompassing overlapping categories and multiple characteristics are classified much less frequently. This type of classification requires a process of recognizing and selecting the core properties of the objects along the lines of function and form, rather than simply shape and colour (English, 2010). Data modelling offers a particularly rich opportunity for classifying irregular and complex objects and, in turn, to develop categorization and statistical reasoning skills. Children must select attributes of interest, attempt to impose structure, and organize and represent their data to communicate and persuade (English, 2012).

11.2.2 Development of Statistical Reasoning

Informal statistical reasoning is the process of developing probabilistic generalizations from data where the claim extends beyond the individual data in question, and the data is used in an explicit way to support the claim (Makar & Rubin, 2009). In reference to young students, Leavy (2008) describes three levels of statistical thinking in young school students when interpreting graphic displays:

- Level 1 (*idiosyncratic*) interpretation includes irrelevant information, shows little awareness of graphing conventions and does not recognize when graphic conventions display the same information.
- Level 2 (*transitional*) recognizes some graphing conventions, through a focus on one aspect of the data only.
- Level 3 (*quantitative*) identifies graphing conventions and can interpret data based upon more than one aspect of the displays. While reasoning at Level 3 is relatively common for students from Grade 3 and above, it is unusual for students below this age (Leavy, 2008).

11.2.3 Studies on Young Children's Modelling

The process of modelling may not a familiar one for young children. English (2012) identified the first stage as selecting attributes of relevance and interest, followed by analysis of a model and subsequent appropriate revisions. Young children's prior experiences rarely include the imposition of structure on items, and children's early observations are typically idiosyncratic collections of ideas rather than groups of features that are selected for inclusion or exclusion.

Lesh and Lehrer (2003) describe modelling as a developmental process with this first stage focused upon observations, followed by decisions regarding which elements are worthy of attention and later a move towards structuring, representing and organizing data. Indeed, applying layers of structure to a data set may be challenging for primary-aged students. Children in Grades 1 and 2 were observed when evolving systems of attributes in a post hoc fashion, for example, but either through lack of opportunity or through limited capacity they did not demonstrate the development of these attributes into a classification system. Even by Grade 5, student models frequently contain redundant or extraneous features (Lehrer & Schauble, 2000).

While many students may not effectively apply structure to data until late in the primary school years, recent research by Kinnear (2013) shows that most children enter school with an intuitive understanding of grouping and with intuitive methods with which to make sense of data. Providing familiar materials such as picture books allowed children to contextualize data discussions and increased their capacity to use data with purpose. With repeated opportunities, even young children can "lift off" common features and start to think abstractly, as documented by English in the context of longitudinal studies of data modelling in the early school years (English, 2010, 2012). Essential to this process is creating a language-rich environment where groups of students have the opportunity to reason verbally both within cooperative groups and with the teacher prior to developing representations (Makar, 2016).

The second stage of model development involves the child developing some type of representation system for communication and revision of the model (English, 2012), described by Lesh and Lehrer (2003) as model documentation. Representations function as more than a record or summary of the child's mathematical conception, because the process of making the representations itself contributes in significant ways to their mathematical and statistical understanding (Thom & McGarvey, 2015). There are many examples of using representations and self-assessments to make sense of data. Mulligan (2015), for example, observed highly capable Year 1 students developing ways of graphical representation which included the use of discrete and continuous data and encompassing a range of meta-representational competencies. By refining their representations over several attempts, students also clarified their reasoning processes.

English (2012) describes the third step of model building as accepting variation within the data set and using the available information for predicting and inferencing. Informal statistical inferencing is the process whereby reasoned, though not statistically tested generalizations are considered and formed through data observations. Key elements include the notion of uncertainty, reliance upon aggregate (not discrete) concepts and acknowledgement that the tendency observed is generalizable beyond the immediate situation.

The final stage of modelling, proposed by Lesh and Lehrer (2003), is the self-assessment principle. In this stage, the modeller determines what did and did not work and in what context: in other words, a process of revision and evaluation. Importantly, this assessment may lead to model refinements and adjustments that would not otherwise be possible (Zawojewski, 2010).

There is an emerging body of evidence indicating that young children have strong, though underdeveloped capacities to participate in model development activities and to use information from their modelling to make decisions or change their previous conceptions regarding a situation. How young students develop and apply the modelling and refinement process is not clearly understood however, especially where the data set is complex or abstract.

11.3 Research Questions

This project focused upon two fundamental questions:

- In what ways do young students develop and apply a model using a complex and undefined data set?
- How do young students use observations and judgements gained from their model to inform their decisions or change previous conceptions about the rules?

11.4 Method

11.4.1 Participants

All students participating in the study were drawn from a Year 1 class (n = 44) in an independent school in metropolitan Sydney, Australia. Selection criteria included high performance in the PASA (Mulligan, Mitchelmore, & Stephanaou, 2015), the Raven's Coloured Progressive Matrices (Raven, Raven, & Court, 1998) and a classroom mathematics assessment. Students who performed in the top ten per cent of their cohort for at least two of the assessments were included.

Nine high-ability students met the selection criteria and were invited to participate in the study. The participating students were aged between 6 years, 4 months and 7 years, 3 months (M = 6 years, 10 months). Students were withdrawn from their regular class to participate in 16 one-hour lessons over one school term with the first author, who is also the gifted and talented teacher at the school.

11.4.2 Procedure

11.4.2.1 Lesson Scope and Activities

The lesson plans loosely followed the structure described by English (2012) namely:

- selecting attributes of relevance and interest for incorporating into a model,
- the development of representational systems for communication and revision of the model, and finally,

- acceptance of variation within the data set, using the available information to make inferences and to make predictions. During this stage, students evaluated the strengths and limitations of their model, although due to their young age and this being their first modelling experience, did not complete Lesh and Lehrer's (2003) final stage of revision and refining.

11.4.2.2 Lesson Phases

Previously, the students participated in four lessons in their Year 1 mathematics class developing skills in data collection around the theme of household pets. This included comparing the numbers of pets of one kind with another, clustering like animals together, and representing these collections on simple picture graphs. Students were given multiple opportunities both individually and in small groups to represent the information in a variety of ways, with some moving towards either clustering information or to regularly spacing and recording information (Fig. 11.1). Students managed to cluster information, present totals or to line up data points showing an early awareness of fundamental data characteristics. These early representations are critical to children's understanding of data, and this process of allowing free pictographs in the early stages allowed children to explore concepts of frequency and variance. Some students were spontaneously including gridlines and totals as shown in other similar studies (Mulligan, Hodge, Mitchelmore, & English, 2013). Consequently, at the start of the lesson sequence, students could define "data" and had some experience of naming, collecting and representing information.

The Year 1 student's pictogram in Fig. 11.1 shows many elements of formal graphing including (1) grouping of totals (top line), (2) division of the space into boxes showing an understanding of equal spacing per unit, (3) marking of "non-data" shown by the use of the X and (4) labelling of columns.

Fig. 11.1 Students' representations of self-collected data about pets

Over the 16 lessons of this design study, students moved through three distinct, though interrelated phases:

- Phase 1: Development of language and representations in preparation for creation of their model.
- Phase 2: Development of rules to distinguish between self-portraits drawn by Kindergarten and Year 3 students (the model).
- Phase 3: Engagement in three reasoning tasks to assess understanding. These were (1) analysing the information about the rules created by the testing to determine the best and worst rule, (2) applying the model in reverse to create illustrations representative of the two age groups and (3) attempting to use the model on a set of self-portraits on Year 2 child artists for the purpose of determining whether these self-portraits were more like Kindergarten or Year 3 self-portraits.

Students worked in self-selected small groups for Phase 1, in teacher-determined groups of three for Phase 2 and individually (with group discussion) for Phase 3.

11.4.3 The Design Process

Using the notion of a hypothetical learning trajectory (Meletiou-Mavrotheris & Paparistodemou, 2015), each stage was pre-planned; however, flexibility both between and within lessons allowed for modifications in pace and focus. Phase 1, for example, was expected to take two lessons; however, this phase actually required six lessons before students were able to move to the formal stage of model building. Thus, while the teacher scaffolded students towards the end goals—in this case the emerging skills of developing reasoning and generalization from data—the teaching process and activities remained learner-responsive.

11.4.4 Phase 1: Classification and Representation of Data

Phase 1 spanned Lessons 1–6 and developed skills in manipulating, describing and representing a collection of data sets. Lessons 1–3 used mathematical manipulatives such as pattern blocks and "quarto" board game pieces to create simple and overlapping sets based upon specific attributes (e.g. colour, height, number of sides). Students then extended their skills by creating overlapping sets of everyday items such as stationery, toys and food supplies. With Venn diagrams constructed using large hoops on tables, the physical manipulation of items allowed students to sort by attributes including shape, function and material. Sets were then described, labelled, photographed and illustrated in students' logbooks. The students were required to choose the attributes of interest and justify why a particular item had

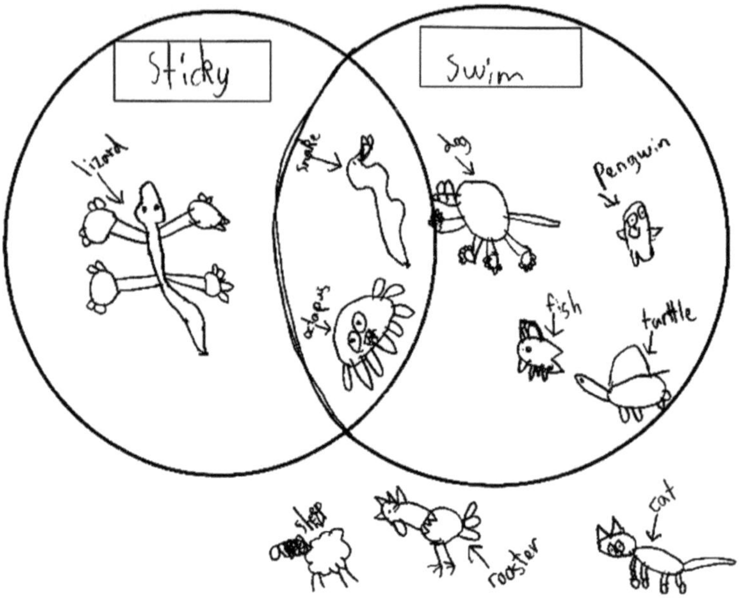

Fig. 11.2 Students' representation of categorization of animals

been classified as belonging to one or both groups. For example, sets of pictures of animals were classified by characteristics including animals that were "sticky" compared with animals that could swim (Fig. 11.2), mammals overlapping with pets and swimming animals overlapping with mammals. Students also completed analogy-type tasks, where they followed rules, seeking exceptions and describing similarities between items.[1]

Figure 11.2 depicts how the students divided into self-selected categories of "sticky" and "swim". Rooster, cat and sheep are considered neither sticky nor able to swim.

11.4.5 Phase 2: Developing a Model by Classifying Self-portraits of Students

11.4.5.1 Orientation to the Task

In Lessons 7 and 8, the rule-based model building was introduced. Teams were each provided a unique set of six self-portraits (illustrations): three produced by Kindergarten and three by Year 3 students. Teams were then set the challenge of

[1]Activities sourced from the Critical Thinking Company (Parks & Black, 1997).

dividing the self-portraits into two groups and providing a set of reasons or rules supporting their decisions. Logbooks and small whiteboards were used to record student comments and observations while the teacher-researcher observed, facilitated conversations and encouraged reflection.

Dividing the self-portraits into year-related groups initially proved challenging: the first classification system proposed by the participating students was by gender, followed by superficial characteristics such as the presence/absence of background or the hair colour of the student artists, or whether the individual was in school uniform or not. (All self-portraits were drawn by individuals at the students' school). Students appeared oblivious to the significant stylistic features of the two age groups. Typical student comments at this stage were "if it is a boy from [school name] it has a green tie", "this picture has no hair" and "Looks like [school name] people they have a full school uniform on". Developing a classification system taking into account the year group of the student artist was dependent upon the teacher-researcher informing the students that some were by young artists and others by older ones. This scaffold allowed students to focus on two age-based categories.

11.4.5.2 Early Attempts at Modelling (Lessons 9–10)

Once student teams could divide the self-portraits into aged-based categories, they moved quite quickly to observe and record similarities within, and differences between, the two groups. Attaching the self-portraits to whiteboards allowed students to move the self-portraits, record observations and document preliminary rules (Fig. 11.3).

Figure 11.3 highlights students' early efforts at separating self-portraits by Kindergarten children from those by Year 3 children. Students' compared specific items (shoes, hair, eyes, etc.); however, the descriptions they used were not yet quantifiable.

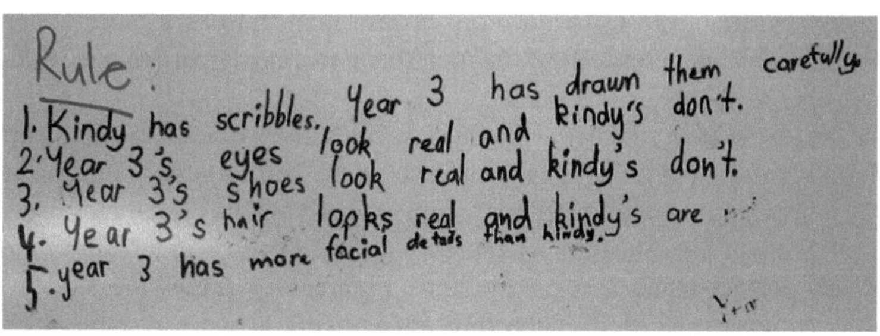

Fig. 11.3 Students' definitions of a set of characteristics

With each rule, the students were encouraged to use specific language and define general terms such as "detailed", "complicated" and "scribble". One team started counting the number of features on the face of each group, a practice subsequently adopted by all teams. A second team noticed differences in eye shapes, eyelashes and feet.

11.4.5.3 Model Development and Testing (Lessons 11–14)

Using the students' observations and rules, teams combined for a teacher-facilitated discussion regarding the most pertinent features to be included in the model. First, a class list was generated of the small-group rules, and then commonalities examined. Class consensus was that the three features noted by all teams included the *shape of hands and fingers*, the *number of features on the face* and the *shape of the eyes*. By encouraging the students to define each rule in terms of properties that could be measured or counted, a set of rules was negotiated as:

- Year 3 self-portraits have thick fingers. Kindergarten ones have no fingers or just sticks.
- If there are nine or fewer features on the face, the illustration is by a Kindergarten student. If ten or more features, then by a Year 3.
- Year 3 self-portraits have eye-shaped eyes and at least one other detail. Kindergarten eyes are balls or dots.

Students were supported in how to count facial features. Essentially if a separate stroke of the pencil was required, it was considered a feature, with the exception of hair, which needed an additional item (i.e. hat, bow, plait) to count as more than 1. A typical Kindergarten self-portrait contained dot eyes ($n = 2$), mouth and perhaps hair or ears ($n = 2$) providing about four to five features in total. The students chose nine features at the limit for Kindergarten illustrations as they seemed to cover a few outliers, while the Year 3 self-portraits contained many more, generally due to the complexity of each attribute requiring many pencil strokes. The students found the rule slow to use because they relied on individual and laborious counting as they were generally unable to estimate more or less than nine as a more experienced person might. Nevertheless, they did not seem to find it particularly difficult to apply.

Once the teams were satisfied with the class rules, they tested these on their unique sets of self-portraits. By this stage, students were comfortable with constructing two-way tables, and mostly found this a convenient way to record their observations (Fig. 11.4 Student classification of representation of each Kindergarten, and each Year 3 self-portrait using a two-way table. Figure 11.4 shows the student representations compared to each of the three rules using a two-way table. Ticks and crosses indicate the student's assessment of compliance. The first "Kindy" illustration, for example, did not comply with Rule 1 or Rule 2; however, it did with Rule 3. In contrast, the first Year 3 self-portrait complied with all three rules. One team's set

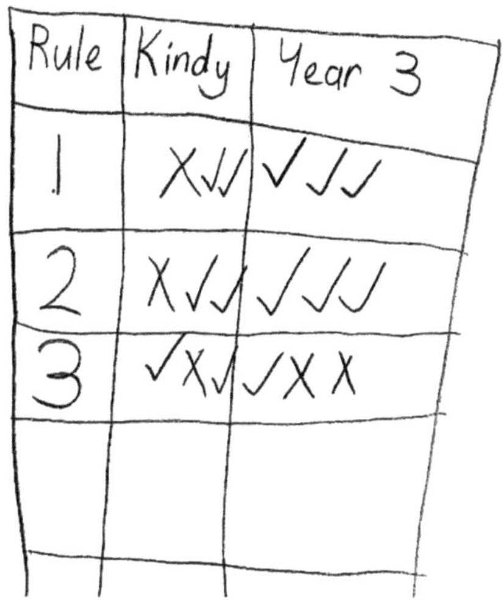

Fig. 11.4 Student classification of representation of each Kindergarten and each Year 3 self-portrait using a two-way table

included several challenging self-portraits, finding they matched only two of the three criteria for its group. Nevertheless, the group accepted the rules as a working model for the next stage of the testing.

Figure 11.4 shows the student representations compared to each of the three rules using a two-way table. Ticks and crosses indicate the student's assessment of compliance. The first "Kindy" illustration, for example, did not comply with Rule 1 or Rule 2, however it did with Rule 3. In contrast, the first Year 3 self-portrait complied with all three rules.

11.4.5.4 Applying the Rules to Two New Data Sets of Self-portraits

Once student teams had developed their models, they tested these on a set of eight new, independent self-portraits. Teams repeated the sorting exercise by firstly dividing the new portraits into apparent age piles and then testing each one against the model rules and recording their observations on new two-way tables with ticks and crosses. Progress was now speedy, and students showed a confidence and efficiency lacking in their previous attempts. This familiarity with the process enabled them to negotiate unexpected problems, for example, when a self-portrait had no hands (Fig. 11.5a), or when the outcome of one rule disagreed with the other two (Fig. 11.5b).

Finally, to confirm understanding, students were provided with an additional 20 self-portraits representing both age groups in mixed order and classified the self-portraits as an individual assessment task. These individual results became a useful resource when supporting their reasoning skills in the following lesson. Following the assessment, students were asked what they regarded to be the best and the worst rule(s).

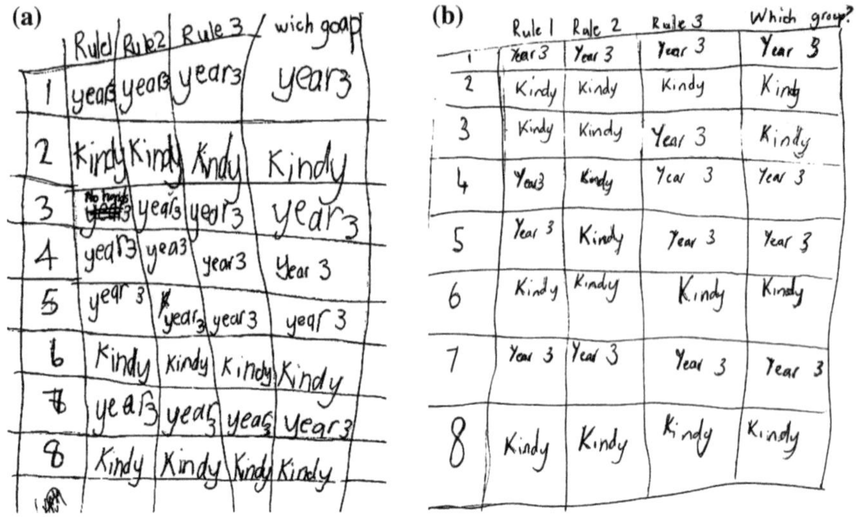

Fig. 11.5 **a, b** Students' categorization of self-portraits by rule using two-way tables

11.4.6 Phase 3: Student's Interpretation of the Model

Phase 3 (Lessons 15 and 16) focused on the analysis of students' reasoning through three new tasks. The first task was for students to determine the best and the worst of the three rules, using the group's categorization of the rules created in the final task of Phase 2 (classifying 20 new self-portraits and applying these to the three rules). The second task involved making two self-portraits, a Kindergarten portrait and a Year 3 portrait, using the three rules. The third required predicting the discriminating power of the model on a new age group of self-portraits, through predicting outcomes, testing self-portraits from the new age group and then revising their predictions in the light of their test results.

11.4.6.1 Task 1—Determining the Best and Worst Rule

At the end of Lesson 14 after students had tested their model on 20 new illustrations, students wrote down what they interpreted as being the best and the worst rule, and how they came to the decision. At the beginning of lesson 15, the teacher-researcher collated the information gained from the students' testing of the model on the 20 new illustrations and presented the results as a clustered column chart (See Fig. 11.6). The researcher discussed the representation of the data, highlighting the six columns presented as three different rules applied separately to the Kindergarten data and the Year 3 data as described earlier in 0. These high-ability students were able to read the graph and understand the function of the key (Rules 1, 2 and 3).

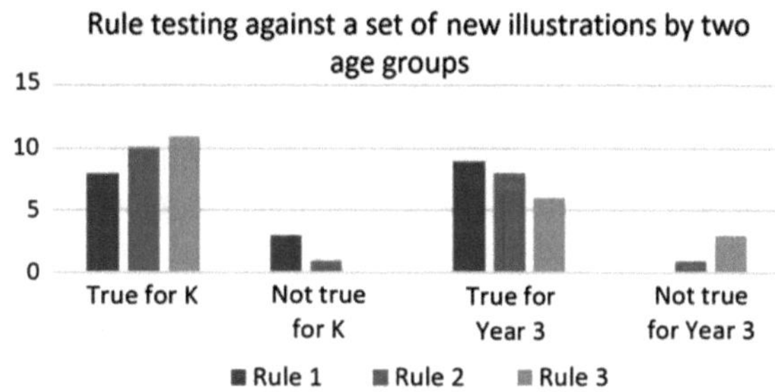

Fig. 11.6 Summary of effectiveness of each of the three rules when used with two age groups

Discussion with the students focused on the effectiveness of the rules when applied to the Kindergarten and Year 3 self-portraits with an emphasis upon which of the rules were most helpful for classification. Students were then directed to review their responses to their initial statements given at the end of Lesson 14 and to compose a new statement. Table 11.1 provides a summary of representative students' statements about the best rule and why, made before (Lesson 14) and after (Lesson 15) interpreting the group graph. Before this interpretation exercise,

Table 11.1 Student responses: Which rule do you think works best? Why?

Level of reasoning (Leavy, 2008)	Lesson 14 Student statements	Lesson 15 Student statements
Idiosyncratic Student 5	Rule two becaus(sic) just 3 year 3 were riten(sic) doan(sic)	The thurd(sic) rule is the best I think becucs(sic) know(sic) Year 3s dident(sic) do the kind'ys (sic) eye shape
Student 9	Rule 1 because Kindy stats [starts] with my name	Rule 2 because it all stars [starts] with my name
Transitional Student 2	Year 3 had thick fingers while Kindy had none or sticks [Rule 1]. I think it was the best because it worked the most	I think rule three is the best because it worked eleven times
Student 3	Rule 2	Rule 2 because the rule is the second-working [second best] in either groups
Student 6	Rule 3 because it doesn't take long to slove [show?]	Rule 3 because It works a lot for kindy and doesen't 9sic) work much for Year 3
Student 7	[Rule] 3 because 9 was Kindy	K 3 (Rule 3) because it worked 100 $ [one hundred %] in our rules in kindy
Student 8	Rule 1 because it worked for most of them	Rule 1 because it worked a lot on Year 3 pictures
Quantitative Student 1	Rule 3 because it almost always worked	Rule 2 because it comes 2nd in our data for both K and for Year 3
Student 4	Rule 1 because I could decide it easily	Rule 2 because rule 1 works least well for kindy but best for Year 3. Rule 3 works best for kindy but least for Year 3, so Rule 2 is the best

Student 9 had a name starting with K (for Kindergarten)

students' responses focused upon their impression of how successful their chosen rule was at differentiating between the self-portraits (Students 1, 2, 3, 7 and 8), or the ease of use (Students 4 and 6). After the exercise, seven of the nine students used the information in the graph to inform their decision regarding the best rule. The level of statistical inference used by students was coded using Leavy's (2008) three levels: *idiosyncratic*, *transitional* and *quantitative*. Individual student responses from the study group were observed at each level, with idiosyncratic thinking from two students (5, 9), transitional thinking observed in five students (2, 3, 6, 7, 8) and quantitative in two (1, 4).

Table 11.2 provides a summary of students' responses to the question of what is the worst rule when asked before and after interpreting the group graph (Lesson 14 cf. Lesson 15). The students' idiosyncratic responses prior to the analysis are important and insightful, particularly for understanding how the students experienced the logistics of using the rules. However, the research focus was on determining whether students chose to incorporate the data if available. After analysis of the graph (Fig. 11.6), five of the nine students were able to provide a data-driven response to the question "What is the worst rule and why?" Similar to the observations recorded in Table 11.1, student responses vary in sophistication with idiosyncratic thinking evident in the responses of three students (5, 6, 9),

Table 11.2 Student responses: Which rule do you think is the worst? Why?

Level of reasoning (Leavy, 2008)	Lesson 14 Student statements	Lesson 15 Student statements
Idiosyncratic Student 5	No response	The secand(sic) because it takes long [to count] and They are dithinkelt [difficult]
Student 6	Rule 1 because it is hard to see some of the fingers in some of the pictures	Rule 2 because it takes a bit to (sic) long
Student 9	No response	Rule 19 (sic) because you can't see it well
Transitional Student 2	If nine or less features on the face its kindy and if its ten or more features on the face [Rule 2]. I think it worked least well because it worked the least	I think rule two worked least well. I think rule two worked least well because it did not work best on any of them
Student 3	No response	Rule 1 because rule 1 is the most non-working rule in kindy
Student 8	Rule 2 because it doesn't work for most of them	Rule 2 because it doesn't work a lot
Quantitative Student 1	Rule 3 because it almost always worked	Rule 2 because it comes 2nd in our data for both K and for Year 3
Student 4	Rule 1 because I could decide it easily	Rule 3 because in kindy it works best in kindy but in Year 3 it worst and is less [well] in Year 3 than Rule 1 in kindy

transitional thinking evident in the responses of four students (2, 3, 7, 8) and quantitative thinking evident in the responses of two students (1, 4).

The students' inferential reasoning is demonstrated through their interpretation of graphical data. Prior to interpreting the graph, student statements focused upon the function of the rule (Students 1, 2, 5, 7 and 8) or the ease of use of the rule (Students 4 and 6). Post-interpretation of the graph, Students 1–8 responded with statements drawing directly upon the class data to validate their decisions. For some students, the response was as general as, for example, Rule 1, because it worked a lot on Year 3 pictures (Student 8). This matched Leavy's (Leavy, 2008) transitional thinking. However, many responses were surprisingly sophisticated. Several students demonstrated quantitative thinking, involving summing up multiple components of the information. Student 4, for example, looked holistically at the data (see Fig. 11.6), responding: "Rule 3 because in kindy it works best but in Year 3 it is worst and it is less [well] in Year 3 than Rule 1 in kindy". This student expresses the conflict in the data itself. Student 1 also recognized the tension in the figures which he resolved by providing two answers, each with a reference population. Both these students were operating at a quantitative level of statistical reasoning, usually seen in students in Grade 3 and above (Leavy, 2008).

There was also a movement from general statements, (e.g. Student 2 "because it worked the most"), to specific data-driven types such as "I think rule 3 is the best because it worked eleven times". The examples documented here of students interpreting, predicting, justifying decisions and exhibiting the flexibility to change their minds demonstrate the impact of their application of the model in new tasks.

11.4.6.2 Task 2—Creating Self-portraits Representing the Two Age Groups

For Task 2, students were required to produce two illustrations: one "like a Kindergarten child might draw" and one "like a Year 3 child might draw". Student responses were compared with each of the three rules of the model (Table 11.3), namely thickness of finger, number of features on the face and the shape of the eye. Shaded squares indicate the student illustration is *not* following the rule. Seven of the nine students produced illustrations distinctly characteristic of the two age groups (Fig. 11.7). Students 2, 3 and 7 all appeared to rework the eyes of the Year 3 drawing, for example, to ensure they followed Rule 3 and their final product contained eyes that were "eye shaped" not "dots or balls. The drawings, as much as the students' reflective comments on the best and worst rule, show their early efforts to articulate the concept of false negatives and false positives and their place in their own critical judgements.

Figure 11.8 shows that while requested to follow the three rules, the student's Year 3 figure meets the requirements of Rule 2 (nine or more features on the face), but does not provide the figure with eye-shaped eyes (Rule 3) or thick fingers (Rule 1).

Table 11.3 Student-generated rule set. Students' successful and unsuccessful (highlighted in bold) attempts to create Kindergarten and Year 3 self-portraits

	Kindergarten			Year 3		
	Rule 1	Rule 2	Rule 3	Rule 1	Rule 2	Rule 3
	Fingers are sticks or absent	Nine or fewer features on the face	Eyes are dots or balls	Fingers are thick	Ten or more features on the face	Eyes are eye-shaped and have at least one other feature
S1	Stick	4	Ball eyes	Thick	10	Oval eyes with pupil
S2	Stick	4	Ball eyes	Thick	24	Semi-circular eyes with eyelashes and pupils
S3	Stick	3	Dot eyes	Puffy	15	Almond-shaped eyes with small pupils and extra line under left eye
S4	Stick	4	Large ball eyes	Thick	13	Almond-shaped eyes with distinct eye balls and pupils
S5	**Thick**	**10**	Circular with pupils and eyebrows	**No fingers**	18	**Ball-shaped with pupils and eye lashes**
S6	Stick	4	Dots	Thick	16	Almond-shaped eyes dot pupils and eyelashes
S7	Stick	7	Circular with pupils	Thick	16	Right eye almond-shaped, lashes and pupil Left eye obscured with large scribble patch
S8	No fingers	6	Large dots	Thick	10	Oval-shaped with dot pupils
S9	No fingers	6	Large balls	**No fingers**	**9**	**Ball-shaped eyes with pupils**

Fig. 11.7 Student illustration showing two representations while following the three rules created by student model

Fig. 11.8 Student illustration showing inconsistent application of the rules

11.4.6.3 Task 3—Using the Model on Self-portraits by Year 2 Students

The final reasoning task for these students was to consider whether Year 2 self-portraits would be most like the previous set of Year 3 self-portraits or most like those done by Kindergarten children. All nine students predicted the Year 2 self-portraits would resemble Year 3 self-portraits more so than Kindergarten ones. Their reasons all related to the relatively small developmental gap between Year 2 and Year 3 which they anticipated would be reflected in the self-portraits produced by the two age groups.

Students were then provided with ten self-portraits by Year 2 artists and asked to measure each one against the model rules, thus deciding which category the illustration matched overall. Because students were working independently, there was some variety in their application of the rules. However, most students found that only two or three of the ten portraits were classified as being like Year 3 self-portraits by the model. The model classified the remaining seven or eight as Kindergarten self-portraits. While most of the Year 2 self-portraits did follow Year 3 self-portraits for Rule 1 (thick rather than stick fingers), Rule 2 (features on face) and Rule 3 (eye shape) identified the self-portraits as similar to Kindergarten illustrations. This would suggest that elongated eye shape at least is a rather late development in children's self-portraits which would explain why Students 2, 3 and 7 took several attempts to create rules following eye shapes for their illustrations described in Table 11.3.

Students were surprised that the Year 2 self-portraits tested as Kindergarten ones using their model. They were asked whether they agreed with their original decision that the portraits were more like Year 3 self-portraits than Kindergarten ones. Of the nine students, seven disagreed with their original decision. Six of these seven provided data-driven inferences by referring back to the information they had just collected.

11.5 Discussion

It is clear that despite their young age, these Year 1 children were capable of formulating an inference based upon documented and reasonable assumptions. They can interact with the content through collection of data and then revisit their original opinion using the new information collected. This demonstration of analytical thinking plus the willingness to change their mind when faced with compelling evidence provides insight into the mathematical capacities of gifted young students.

In this study, the researchers aimed to observe the ways in which young students developed and then used a model using a complex and undefined data set. The use of self-portraits for model construction has precedence, with Lehrer and Schauble (2000) also using portraits to support students in developing classification skills. As in Lehrer and Schauble's (2000) study, the researchers found an initial preoccupation with superficial features and ad hoc observations. The students in the present study showed an even stronger trajectory of growth, however.

As they progressed through their modelling activities, Lehrer and Schauble's youngest learners developed models that were essentially a collection of descriptors encompassing the self-portraits. In contrast, the students in this study were able to develop and apply an entirely rule-based model. Given that students in Lehrer and Schauble's study were of mixed ability, this difference in findings is likely to reflect the capacity of the Year 1 highly able students to reason and think abstractly at a younger age than peers (Harrison, 2005). One of the students (Student 9), although able to participate willingly in the class activities, was unable at this stage to use data to inform decision-making or to transfer an understanding of the rule into personal illustrations. It is suggested that the development shown by this student is a more typical example of results for this age group than that produced by the majority of the other students in this study and broadly consistent with Lehrer and Schauble's group described above.

As in Phase 1 of the study, English (2010) found that young students made decisions about which pieces of information were relevant or irrelevant, displaying meta-representational knowledge in structuring their classification of data. Makar

(2016) also studied inferential thinking in young children and selected articulate and highly able students. As in this study, Makar devoted the early learning sequence to foundational skills for predicting, checking and developing inferential reasoning. The model construction process was effectively developed through explaining, justifying and representing thinking, all of which formed essential phases in the Year 1 project described here.

Modelling can be viewed as a cyclic activity with the transitional stages involving a self-assessment of the model (Lesh & Lehrer, 2003) followed by a refinement process (Zawojewski, 2010). The Year 1 students developed this refinement process through their individual judgements regarding the strengths and weakness of the three model components. They recognized that no model for something as variable as children's self-portraits could guarantee accuracy, but they were prepared to continue modelling despite uncertainty. However, they did understand the rules did not work equally well on all age groups, and their model was particularly ineffective when discriminating between the existing Kindergarten age group and the new Year 2 one. Lehrer and Schauble (2000) support this process in that they assert that comparison of two or more models is the next stage in the modelling cycle where students compare each for their relative fit for the "world".

The children in the present study used two "worlds"—Year 2 and Year 3 self-portraits in their modelling and found that the model developed for the Year 3 "world" did not allow for discrimination between self-portraits in the "Year 2" world. While students modified their thinking about the rules, they did not move on to the next step and suggest changing the model itself or adapting it in any way. Such a step may have produced a model that would discriminate between the Year 2 self-portraits and from those in Kindergarten.

11.6 Conclusion and Implications

This study focused on a small sample of highly able students from one school, and such observations are not generalizable. However, the findings are consistent with an emerging number of studies such as those of English (2010, 2012, 2013), Kinnear (2013), Makar (2016) and Mulligan (2015) which demonstrate that students in the early years of school can analyse complex data sets and develop rule-based models; they do so through organizing and representing, justifying and adapting their models.

Students did reflect to a large extent the phases outlined by Lesh and Lehrer (2003), albeit with more able students. Students developed and refined their model through small group discussion, trial and error and then tested it on new data sets. Key to the success of the model was the students' capacity to move beyond ad hoc observations by pooling observations and identifying key elements (e.g. the number of features, the shapes of hands and eyes). By developing rules based upon measures that could be specifically described and counted, the resulting model was robust enough to discriminate on several sets of test self-portraits. Having clear rules involving shapes and numbers made the application of the model relatively straightforward. Moreover, as observed by the teacher-researcher, the students operated confidently and independently and could use data obtained to predict outcomes for other cases. When the new cases did not match their predictions, students were able to use inferential reasoning to revisit and alter previous decisions.

The demonstrated achievements for these students in design, representation, testing, modifying and using reasoning skills would suggest that for some students at least their capacities are underestimated in the regular Year 1 classroom. The level to which children were prepared to change their rule based on the evidence is another example of students' reasoning being above age predictions in regard to developmental expectations (Lesh & Carmona, 2003). Incorporating modelling activities into the early years of schooling to develop rule-based models and reasoning skills is rarely prioritized but it is critical to advancing students' critical numeracy capabilities. This study demonstrates a more pressing need for professionals to acknowledge the benefits of this approach for most students and provide appropriately paced and scaffolded opportunities. This demonstration of analytical thinking as well as the students' willingness to change their mind when faced with compelling evidence provides new insight into the mathematical capacities of gifted young students.

References

Clements, D., & Sarama, J. (2009). *Learning and teaching early math: The learning trajectories approach*. New York, NY: Routledge.

Diezmann, C. M., & Watters, J. (1998). Thinking by young children during argumentation: Use of evidence and logic. In *Thinking processes: Going beyond the surface curriculum* (pp. 115–134).

English, L. (2010). Young children's early modelling with data. *Mathematics Education Research Journal, 22*(2), 24–47. doi:10.1007/BF03217564.

English, L. (2012). Data modelling with first-grade students. *Educational Studies in Mathematics, 81*(1), 15–30.

English, L. (2013). Reconceptualizing statistical learning in the early years. In L. English & J. Mulligan (Eds.), *Reconceptualizing early mathematics learning* (pp. 67–82). New York: Springer.

Gopnik, A. (2012). Scientific thinking in young children: Theoretical advances, empirical research, and policy implications. *Science, 337,* 1623–1627.

Harrison, C. (2005). *Young gifted children: Their search for complexity and connection*. Exeter, NSW: Inscript Publishing.

Inhelder, B., & Piaget, J. (1958). *The growth of logical thinking from childhood to adolescence: An essay on the construction of formal operational structures*. New York: Basic Books.

Kinnear, V. (2013). Young children's statistical reasoning: A tale of two contexts. *Ph.D.* Queensland University of Technology, Brisbane, QLD.

Leavy, A. (2008). An examination of the role of statistical investigation in supporting the development of young children's statistical reasoning. In O. N. Saracho & B. Spodek (Eds.), *Contemporary perspectives on mathematics in early childhood education* (pp. 215–232). Charlotte, NC: Information Age Publishing.

Lehrer, R., & Schauble, L. (2000). Inventing data structures for representational purposes: Elementary grade students' classification models. *Mathematical Thinking and Learning, 1–2,* 51–74. doi:10.1207/S15327833MTL0202_3.

Lesh, R. A., & Carmona, G. (2003). Piagetian conceptual systems and models for mathematizing everyday experiences. In R. Lesh & H. M. Doerr (Eds.), *Beyond constructivism: Models and modeling perspectives on mathematics problem solving, learning, and teaching* (pp. 71–96). Mahwah, N.J.: Lawrence Erlbaum Associates.

Lesh, R. A., & Lehrer, R. (2003). Models and modeling perspectives on the development of students and teachers. *Mathematical Thinking and Learning, 5,* 109–129.

Makar, K. (2016). Developing young children's emergent inferential practices in statistics. *Mathematical Thinking and Learning, 18*(1), 1–24. doi:10.1080/10986065.2016.1107820.

Makar, K., & Rubin, A. (2009). A framework for thinking about informal statistical inference. *Statistics Education Research Journal, 8*(1), 82–105.

Meletiou-Mavrotheris, M., & Paparistodemou, E. (2015). Developing students' reasoning about samples and sampling in the context of informal inferences. *Educational Studies in Mathematics, 88*(3), 385–404. doi:10.1007/S10649-014-9551-5.

Morsanyi, K., Divine, A., Nobes, A., & Szucs, D. (2013). The link between logic, mathematics and imagination: Evidence from children with developmental dyscalculia and mathematically gifted children. *Developmental Science, 16*(4), 542–553. doi:10.1111/desc.12048.

Mulligan, J. (2015). Moving beyond basic numeracy: Data modeling in the early years of schooling. *ZDM Mathematics Education, 47,* 653–663.

Mulligan, J., Hodge, K., Mitchelmore, M., & English, L. (2013). Tracking structural development through data modelling in highly able grade 1 students. In V. Steinle, L. Ball, & C. Bardini (Eds.), *Proceeding of the 36th annual conference of the mathematics education research group of Australasia* (pp. 530–536). Melbourne, Vic: MERGA.

Mulligan, J., Mitchelmore, M., & Stephanaou, A. (2015). Pattern and Structure Assessment: an assessment program for early mathematics (F-2) Response Booklets F. 1, 2. Camperdown, VIC: Australian Council for Educational Research.

Parks, S., & Black, H. (1997). *Building thinking skills® book 1.* Pacific Grove, CA: Critical Thinking Company.

Raven, J., Raven, J. C., & Court, J. H. (1998). *Manual for Raven's progressive matrices and vocabulary scales* (Vol. 2). San Antonio, TX: Pearson.

Thom, J. S., & McGarvey, L. M. (2015). The act and artefact of drawing(s): Observing geometric thinking with, in, and through children's drawings. *ZDM: The International Journal on Mathematics Education, 47*(3), 465–481. doi:10.1007/s11858-015-0697-0.

Zawojcwski, J. S. (2010). Problem solving versus modeling. In R. A. Lesh, P. L. Galbraith, C. R. Haines, & A. Hurford, *Modeling students' mathematical modeling competencies* (pp. 237–244). New York: Springer. doi:10.1007/978-1-4419-0561-1.

Authors Biography

Dr. Gabrielle Oslington is gifted and talented coordinator at an independent school in Sydney, and she is currently completing a Ph.D. in Education at Macquarie University on mathematical reasoning in young gifted mathematicians. She has interest in the identification and support of mathematically gifted students particularly in the early years. Her study is contextualized within her school where she has responsibility for the students whose academic ability places their needs beyond the normal differentiation provided by the classroom teacher and works with children across their primary schooling. Dr. Oslington has a particular interest in the needs of twice exceptional children and has postgraduate training in children with autism spectrum disorders.

Dr. Joanne Mulligan is a Professor of Education at Macquarie University specializing in mathematics and science education from early childhood to secondary level. Dr. Mulligan's research focuses on the cognitive/psychological aspects of learning mathematical concepts, including investigating children's underlying awareness of pattern and structural relationships and spatial reasoning. Dr. Mulligan has developed and evaluated new assessment instruments and pedagogical programmes, the Pattern and Structure Assessment (PASA) and the Pattern and Structure Awareness Programme (PASMAP), in collaboration with the Australian Council for Educational Research (ACER). She is currently project leader of the Opening Real Science Project, a collaborative cross-institutional project focused on improving primary and secondary teacher education programmes.

Dr. Penny Van Bergen is a Senior Lecturer in Educational Psychology in the Department of Educational Studies at Macquarie University. She has a background in educational and developmental psychology, with a focus on children's development of cognitive and socio-emotional skills. Dr. Van Bergen graduated with a Ph.D. in Psychology from the University of New South Wales in 2008. In her research, Dr. Van Bergen has explored children's recall of salient emotional events, how children's memory formation is linked to emotional understanding and perspective taking and how parents and teachers contribute to memory. Dr. Van Bergen is also passionate about enhancing the research opportunities of education students.

Chapter 12
Making Connections to Realize Learning Potential in Early Childhood Mathematics

Aubrey H. Wang and James P. Byrnes

Abstract In this chapter, we discussed how the Opportunity-Propensity (O-P) framework could be used to conceptualize and test the effects that children's (0–8 years) background, mathematics learning opportunities, self-regulation, and prior achievement have on mathematics learning. In prior studies of the O-P framework, we identified and verified the predictive role of antecedent factors such as family socioeconomic status, parent educational expectations for their children, age, birth weight, gender, and ethnicity. With respect to opportunity factors, we identified and confirmed the predictive role of several aspects of instruction. With respect to propensity factors, we have identified and confirmed the role of prior knowledge, motivation, and self-regulation. We encourage our international researchers to build on this work in order to create the most accurate predictive model of early children mathematics achievement. This way, we can collaborate in guiding early mathematics policymakers, practitioners, and professional and advocacy organizations by providing them with a framework on early mathematics achievement to scaffold understanding, generate and test hypotheses, and adapt targeted interventions to address context- and cultural-specific problems.

Keywords Opportunity-Propensity framework · Opportunity factors
Propensity factors · Structural equation modeling · Mathematics learning
Early childhood

A. H. Wang (✉)
Saint Joseph's University, Philadelphia, USA
e-mail: awang@sju.edu

J. P. Byrnes
Temple University, Philadelphia, USA
e-mail: jpbyrnes@temple.edu

V. Kinnear et al. (eds.), *Forging Connections in Early Mathematics Teaching and Learning*, Early Mathematics Learning and Development,
https://doi.org/10.1007/978-981-10-7153-9_12

12.1 Background

In the 1990s and early 2000s, retrospective publications began to appear, suggesting that most studies conducted by educational researchers were not particularly scientific nor focused on solving real educational problems in the USA (Kaestle, 1993; Shavelson & Towne, 2002; Walters & Lareau, 2009). Although this claim was disputed by some (e.g., Walters & Lareau, 2009), it ultimately affected the conduct of subsequent research due to two key factors. The first was a report of the National Research Council (NRC) of the USA that focused on what it means for educational research to be scientific (Shavelson & Towne, 2002). Special issues of journals and texts were devoted to reacting to this influential report.

The second source of change was the fact that the Institute of Education Sciences (the primary grant funding agency for educational research in the USA) adopted many of the assumptions of the NRC publication and created a new paradigm for making decisions about the kinds of studies that it would fund. This paradigm assumed that scientific designs in education can be arrayed along a progression ranging from (1) designs that simply reveal phenomena or outcomes that need to be explained (exploratory studies such as surveys) to (2) designs that reveal the correlates or predictors of these outcomes as a means of building theories or explanations of these outcomes (e.g., multivariate longitudinal studies) to (3) finally experimental designs that create interventions that target the causal factors identified by the constructed theories (theory testing designs). The goal is to be always progressing toward creating theories and theory-based interventions in this new view.

12.2 Our Approach to Theory Building

Our approach to theory building relies on several strategies to efficiently identify the predictive factors that could serve as the core constructs of a multivariate theory that can explain achievement disparities between individuals, groups, or schools. The first strategy is to build a comprehensive, multivariate theory out of smaller theoretical accounts that have consistent empirical support but appeal to only one or two factors. With a few notable exceptions, most educational fields, including the field of educational psychology and mathematics education, are unfortunately divided into specialist camps of researchers who primarily focus on one construct (e.g., motivation) and not on the constructs that are important to researchers in other camps (e.g., study strategies).

To fully account for achievement disparities, it would seem that theorists need to begin to incorporate constructs from these multiple camps into a single, more comprehensive account. The second strategy is a derivative consequence of taking a more comprehensive, multivariate approach: When more predictors are included in a model in a longitudinal study, one can see which predictors account for the most variance in the outcome variable (therefore, more predictive) and also see whether some predictors turn out to be spurious because prior less comprehensive studies failed to include the right mix of powerful, authentic predictors in their designs.

The third strategy is to test the comprehensive model using an existing, national database that (a) includes information on a number of potential predictors of achievement outcomes and (b) followed a large number of participants longitudinally. This strategy is obviously more efficient and cost-effective than the more common approach in which individual groups of researchers collect less comprehensive data from single schools. Grant agencies in the USA recognized the utility of funding these large-scale multivariate studies in order to facilitate the discovery and theory-building process. Certainly, this belief is shared by government agencies and early education researchers around the world as they have funded large-scale multivariate studies such as the Longitudinal Study of Australian Children, the Québec Longitudinal Study of Child Development for the same purpose. The fourth strategy is to use advanced quantitative methods such as structural equation modeling to improve our measurement, hypotheses, and relationships between predictors and outcomes.

In the last section of this chapter, we propose a fifth strategy to increase the generalizability of the model by asking our international colleagues to replicate and extend the model using similar longitudinal studies drawn from different national populations of children and their communities. We believe this proposed strategy will generate strong evidence to support the generalizability of our theory across national contexts in the following ways: (1) Understand how well the model fits data collected from different populations of children; (2) examine which core variables remain as consistent predictors and which become spurious; and (3) investigate new variables that can be added to further strengthen the model. Our long-term goal is to create a web that would house the accumulated research on the OP model of achievement (Byrnes, 2003; Byrnes & Miller, 2007; Byrnes & Wasik, 2009; Jones & Byrnes, 2006; Wang, Shen, & Byrnes, 2013); highlight the simplicity and flexibility of the O-P model for use in research, policymaking, and program implementation; and connect the growing number of O-P researchers and users around the world.

12.3 Opportunity-Propensity Model

The present study benefits from the previous theory-building efforts of researchers who have conducted both secondary analyses and prospective smaller studies in order to test and the O-P model of achievement (Byrnes, 2003; Byrnes & Miller, 2007; Byrnes & Wasik, 2009; Jones & Byrnes, 2006; Wang et al., 2013). The multivariate O-P model is an attempt to synthesize existing theoretical models of psychological constructs that have been individually found to be associated with literacy, science, and math achievement. In all studies, academic achievement was measured by standardized assessments. For the studies on early mathematics achievement (e.g., Byrnes & Wasik, 2009; Wang et al., 2013), the items were designed by experts in early childhood development to measure children's conceptual knowledge, procedural knowledge, and problem-solving skills. Test items related to number sense, number properties, operations, geometry and spatial sense, data analysis, statistics, probability, patterns, and algebra, and functions.

The psychological constructs include domain-specific (prior) knowledge (Hailikari et al., 2008; Murphy & Alexander, 2002), motivation (Wigfield & Cambria, 2010), intelligence (Soares et al., 2015), and self-regulation (Schunk & Zimmerman, 2013). But the model also extends beyond these psychological accounts by attempting to integrate this work with scholarship in other fields such as early childhood education, mathematics education, and education policy where researchers have likewise been interested in achievement but focused on constructs not ordinarily investigated by educational psychologists such as formal and informal mathematics, effective teaching methods, and the notions of educational opportunity and income-based educational disparities. Furthermore, we are using advanced quantitative methods such as structural equation modeling for theory building.

12.4 Opportunity and Propensity Conditions

The model building began by asking the core question, "Why do some children attain higher levels of achievement on end-of-year tests than other children?" The answer that was ultimately generated through an intensive, synthesis-oriented examination of the literatures in educational psychology, teacher education, and education policy was that (a) higher scoring children were presented with more high-quality opportunities to learn the content on these tests than lower scoring children and (b) higher scoring children were more willing and able to learn this content when it was presented.

The former requirement was dubbed the "opportunity condition," and the latter was dubbed the "propensity condition" (e.g., Byrnes & Miller, 2007). Both conditions have to be fulfilled in order to obtain high achievement. Typically, educational psychologists tend to focus more on propensity factors than opportunity factors (because they focus on characteristics of learners more than characteristics of high-quality instruction), and mathematics education and education policy researchers tend to focus on opportunity factors more than propensity factors (because they focus on high-quality instruction and learning opportunities more than characteristics of learners).

One merely needs to read standard textbooks in educational psychology, mathematics education, or education policy to see the relative weight given to opportunities and propensities in these distinct fields. The model-building efforts of O-P theorists conducted since 2005 were oriented toward identifying variables that should be included in the two categories of opportunity factors and propensity factors. That is, many claims have been made about instructional and child factors that predict achievement, but it has not all been clear which would continue to predict after all (or most) are included in the same multivariate longitudinal study.

This process of distinguishing between authentic predictors and spurious predictors is an essential first step in building an accurate multivariate theory and eliminating factors from consideration. So far, secondary analyses of the National Assessment of Educational Progress (NAEP), National Educational Longitudinal Study of 1988 (NELS:88), and Early Childhood Longitudinal Study-Kindergarten (ECLS-K) and Early Childhood Longitudinal Study-Birth (ECLS-B) databases have shown that the propensity factors that have remained significant predictors after multiple controls include prior domain-specific knowledge, domain-specific interest/competence, and self-regulation.

The opportunity factors that have remained significant after controls include content exposure (indexed by courses taken and teacher reports of content coverage) and teacher-reported style of presentation (traditional, reform, and balanced approaches in literacy, science, and math). As we report later, there are also factors in a third category called antecedent factors, but their role can best be understood after describing propensity factors and opportunity factors in a little more detail.

12.5 Propensity Factors

As noted above, disparate groups of researchers in educational and developmental psychology examined factors such as preexisting knowledge, motivation, and self-regulation in relative isolation from each other. OP theorists have endeavored to integrate these separate strands of research into a single model. The first step in doing so was to identify characteristics of learners that are consistent predictors of achievement by reviewing the literature (e.g., Byrnes & Miller, 2007). The second step was to understand how and why these variables might explain different levels of knowledge growth in children across an academic year. In what follows, we briefly discuss the conclusions of this second step.

Before doing so, however, we should note that "achievement" is typically equated with scores on high-stakes standardized tests in many of the studies that we report. In the USA and elsewhere, teachers are held accountable for presenting specific content that is assessed on these standardized tests. Our goal was to develop a theory that explains why some children seem to acquire more of this content than other children in order to help inform intervention and policy efforts. But the theory could explain the acquisition of any kind of knowledge or skill (e.g., in music, visual arts) not just those assessed on standardized tests. We also recognize that standardized tests also privilege certain cultural groups over others and also often sometime require procedural skill more than conceptual understanding, but large-scale data sets such as the National Assessment of Educational Progress (NAEP) and Early Childhood Longitudinal Study-Kindergarten sample (ECLS-K) require computational skill, conceptual understanding, and problem-solving.

Prior Achievement. The strongest predictor of later achievement in a domain (e.g., math) is prior achievement in that domain, followed by self-regulation and motivation (Byrnes, 2011; Byrnes & Miller, 2007; Byrnes & Wasik, 2009; DiPerna et al., 2005; McClelland et al., 2007).

IQ. General ability (e.g., IQ) also predicts but tends to explain less than 10% of the variance of the dependent variable or is less predictive when prior knowledge, self-regulation, and motivation are controlled (Sternberg et al., 2001).

Motivation and Self-Regulation. Research has shown that children's motivation (including their goals, interests, and self-efficacy) and self-regulation skills and academic achievement are significantly related. Consistent evidence suggests that greater levels of attention, task persistence, and active participation have strong associations with standardized test scores and teacher-rated achievement that is independent of initial cognitive ability and prior basic skills (Alexander et al., 1993; DiPerna et al., 2005; Duncan et al., 2007; Hindman et al., 2010).

12.6 Opportunity Factors

As for the explanatory basis of the opportunity factors confirmed to date through our extensive review of the literature to identify consistent predictors, the role of content exposure is relatively self-evident: Students cannot be expected to show mastery of content that was not presented by their teachers. Thus, measures of whether content was exposed should predict achievement, especially when these measures are accurate and precise. Even though the year preceding kindergarten has been found to be extremely important in mathematics development (Clements & Sarama, 2009; NAEYC, 2002), results from the few observational studies of prekindergarten teachers and programs in the USA suggest that in general, very little mathematics is normally presented during the prekindergarten years (Early et al., 2005; Graham et al., 1997; Lamy et al., 2004; Clements & Sarama, 2007, 2008; Tudge & Doucet, 2004). This finding was revealed by evaluation studies of 14 US-based prekindergarten (for 3- and 4-year-olds) curriculums including Bright Beginnings, Creative Curriculum, and others which found that most of these programs were built on literacy goals with minimal time devoted to mathematics (Farran et al., 2007).

In response, a number of researchers conducted experimental studies to examine the effects of structured early mathematics curriculum on early mathematics knowledge and skill of prekindergarten children from low-income families (Chard et al., 2008; Clarke et al., 2011; Clements & Sarama, 2007, 2008; Clements et al., 2011; Starkey et al., 2004). As for the relative weight given to mathematics content and reading skills in early childhood programs in countries other than the USA, we are unaware of comparable studies. In international studies of older children (e.g., PISA, TIMSS), nations do differ in content exposure and there is a corresponding difference in achievement levels.

12.6.1 Early Mathematics Curriculum

Wang, Firmender, Power, and Byrnes (2016) recently conducted meta-analysis of 29 experimental and quasi-experimental studies of early mathematics programs for prekindergarten (for 3- and 4-year-olds) and kindergarten (age 5–6) environments. They found an overall moderate-to-large average effect size (Cohen's $d = 0.62$, range of 0.50–0.75) across these 29 early mathematics programs. The 10 included studies that evaluated four mathematics curricula, Building Blocks Curriculum, Early Learning in Mathematics, Experimental Mathematics Curriculum, Pre-K Mathematics Curriculum, had a moderate-to-large average effect size (Cohen's $d = 0.63$, range of 0.44–0.82).

The mode minutes of mathematics exposure across these curricula was 63 min per week and 1450 total minutes across the whole curriculum (Wang et al., 2016). For example, the Building Blocks' instructional approach is finding the mathematics in, and developing mathematics from, children's activity (Clements & Sarama, 2007). Children are guided to extend and mathematize (i.e., explicate, articulate, and describe) their everyday activities, from block building to art to songs to puzzles, in mathematical language. Thus, the processes of communicating and reasoning, and mathematizing are continually developed through discussions.

Activities include whole group (about 10 min per day), small group (10–15 min once per week for each child, working in groups of four with the teacher), and centers (including a computer center, 5–10 min twice a week for each child). The curriculum includes 30 weeks of instruction; teachers completed from 24 to 30 weeks. Teachers ask students to solve problems or tasks and then ask such questions as "How do you know?", "Why?", and "Can you tell how you figured that out?" More detailed descriptions of Building Blocks are available (Clements & Sarama, 2007, 2008; Clements et al., 2011).

12.6.2 Supplemental Mathematics-Related Activities

In the Wang et al. (2016) meta-analysis, studies that were coded as supplemental mathematics-related activities (n = 19), or interventions that were implemented in addition to the regular mathematics curriculum, were found to also have a moderate to large average effect size (Cohen's $d = 0.63$, range of 0.45–0.81). The mode minutes of mathematics exposure across these studies was 90 min per week and 720 min in total (Wang et al., 2016). For example, Ramani and Siegler (2008, 2011) engaged young children to play linear board games as a way to develop their numerical representations. An experimenter met individually with each preschooler for five 15- to 20-min sessions within a three-week period. Sessions were held in either their classroom or an unoccupied room nearby. The experimenter used The Great Race linear board game. It included a board that was 52 cm wide and 24 cm high with the name of the game printed at the top of the board. Below the name were 10 equal-sized squares of different colors arranged in a horizontal array.

Each square contained one number, with the numerical magnitudes increasing from left to right. Experimenters asked children to play a game that developed their knowledge and skills in counting, number line estimation, magnitude comparison, numeral identification, and arithmetic. Experimenters asked such questions as: "If this is where 0 goes (pointing) and this is where 10 goes (pointing), where does N (e.g., 5) go?", "Can you choose the bigger number between these two numbers?", "Can you name the numeral on this card?", and "Suppose you have N oranges and I give you M more; how many oranges would you have then?" More detailed descriptions of the linear board game are available (Ramani & Siegler, 2008, 2011; Siegler & Ramani, 2009).

Moreover, there are emerging evidence suggesting that children from low-income families have differential exposure to even the typical mathematics learning opportunities during the early years (Wang, 2010) and that low-income kindergarten children's exposure to analytic and reasoning mathematics activities is significantly related to their mathematics test scores (Georges, 2009; Wang, 2010). For instance, Georges (2009) found that while instruction explained only four percent of the variance in mathematics test scores attributable to classrooms for all kindergarten students, the portion of variance attributable to classrooms was larger (or more predictive) for students in high-poverty classrooms than in low-poverty classrooms. She further found that for high-poverty classrooms, activities that built on students' analytic and reasoning abilities and worksheet-related activities were positively related to students' test scores in all subtests (Georges, 2009). However, Wang (2010) found that there were statistically significant variations in the frequency in which children in poverty engaged in analytic and reasoning activities, suggesting that this group of children has differential exposure to typical, early mathematics learning opportunities.

12.7 Antecedent Factors

As noted above, variables in the two categories of opportunity factors and propensity factors do not account for all of the variables that have been found to be predictive of individual and group differences in early mathematics achievement. Recall that opportunity and propensity factors provide answers to the initial question, "Why do some children demonstrate higher levels of conceptual and procedural knowledge of mathematics than other children?" (answer: The former were provided with more opportunities to learn mathematics and were more prone to take advantage of these opportunities).

The third set of factors emerged when a second-tier question is asked: Why are some children presented with more opportunities to learn in early childhood and enter these opportunities more willing and able to learn math? Analysis of literatures beyond educational psychology and teacher education revealed that there are aspects of children's home lives and sociocultural demographics that have also been found to be predictive of achievement and need to be incorporated into whatever comprehensive, synthetic model is ultimately constructed.

Because factors in the third occur earlier in time than opportunity and propensity factors and help explain their emergence, O-P theories call them antecedent factors. They include age, birth weight, parental expectations, maternal education, and family income.

12.7.1 Age

Age has a well-established relationship to children's mathematical competence (Jordan et al., 2006; Hindman et al., 2010; Ransdell & Hecht, 2003). Age is appropriately construed as an antecedent factor because it can explain why some children were exposed to more opportunities than others (e.g., older children usually have been exposed to more opportunities than younger children) and why some children might be more prone to take advantage of these opportunities (e.g., an increase in cognitive skills due to brain maturation).

Generally, very young children begin with basic mathematics competence that over time develops into more complex mathematical skills. This is especially true during the preschool ages, when children's mathematical competence develops from recognizing small groups of objects to counting the number of objects in correct order and, eventually, developing early arithmetic skills (Geary, 2007). Furthermore, age is an educationally relevant variable since there is an 11- to 12-month variation in children's chronological age during each academic year of schooling (Dowker, 2008).

12.7.2 Birth Weight

Low birth weight (born less than 2,499 g), which is prevalent among low-income children (Collins & David, 1990), has been associated with greater likelihood of increased cognitive delay at 2 years of age (Hillemeier et al., 2010), increased frequency of impaired language functioning (Sajaniemi et al., 2001), lower scores on standardized measures of academic achievement (Bowen et al., 2002), and greater likelihood to fall behind academically (Bowen et al., 2002).

12.7.3 Parental Expectations

Higher parental expectations for children have been associated with greater likelihood of selection of more core academic courses (Catsambis, 2001), better school attendance (Kurdek & Sinclair, 1988), and stronger academic performance (Fehrmann et al., 1987; Rutchick et al., 2009). Even with socioeconomic status controlled, parental expectations have been found to explain significant variance in opportunities, propensities, and achievement (Byrnes & Miller, 2007; Byrnes & Wasik, 2009).

Parental expectations have been found to influence child expectations (Rutchick et al., 2009) and motivation (Wood et al., 2011), both of which are associated with better academic performance.

12.7.4 Maternal Education

Maternal education has been found to have direct and indirect, positive relationships with achievement (Davis-Kean, 2005; Eccles, 2005) and explains much of the variance in cognitive outcomes for low- income prekindergarten children (Perry & Fantuzzo, 2010). Maternal education has been associated with greater exposure to literacy and numeracy skills at home (Lung et al., 2011), more educational opportunities in local communities (Furstenberg et al., 1999), and higher level of schooling expected of their children (Byrnes & Wasik, 2009; Davis-Kean, 2005), all three of which are predictive of stronger academic outcomes. Additionally, maternal education has been found to indirectly affect a child's academic achievement to the extent that it influences family income and residence (Coleman, 1987; Furstenberg et al., 1999).

12.7.5 Family Income

Research in the USA has clearly documented the existence of achievement differences by family income. A seminal research that empirically examined a nationally representative sample of kindergarten children found substantial differences in children's test scores by family income, with children in the highest income group scoring 60% above the scores of children from the lowest income group as they begin kindergarten (Lee & Burkam, 2002). A more recent study utilizing data from random assignment studies on welfare and poverty programs in the USA found that a $1,000 increase in annual income increases young children's achievement by 5–6% of a standard deviation (Duncan et al., 2011).

12.8 Using Structural Equation Modeling for Theory Building

Structural equation modeling (SEM) provides mathematics education and early childhood researchers and practitioners with the ability to estimate multiple cross-dependency relationships and capture unobserved variables in such relationships while controlling for measurement error (Bollen, 2011). SEM, also known as causal modeling, simultaneous equation modeling, covariance and mean structure modeling, is used to confirm the developed theoretical framework and offers multiple ways to evaluate the validity of the model (Bollen, 2011). Typically, SEM proceeds in two stages. The first stage is the estimation of the measurement model, which represents a set of observable variables as multiple indicators of a smaller set of latent variables (dimensions of opportunity, propensity, and antecedent factors), and the theoretical model, which describes dependency relations between the latent variables, and which is grounded in theory (hypotheses). The second stage is the estimation of the structural model, which is a combination of the measurement and path models.

For instance, in our 2013 study (Wang et al., 2013), we tested the O-P model on mathematics achievement of a subsample of 350 African American, 350 Hispanic American, and 300 White children who were below poverty line from the restricted use Early Childhood Longitudinal Study-Birth (ECLS-B) database. As a first step, we used measurement models to consolidate the multi-dimensional latent constructs of antecedent, opportunity, and propensity. Our antecedent factor was indicated by birth weight, early cognition, age, parent expectation, and mother's years of education.

Opportunity was indicated by the latent constructs of teacher-initiated activities for learning basic mathematics skills and teacher-initiated, integrated learning activities. Propensity was indicated by the latent constructs of prekindergarten parent rating and prekindergarten provider rating, as well as preexisting cognition. Prekindergarten parent and provider ratings were indicated by items from previous research. A confirmatory factor analysis, as part of the measurement model (see Fig. 12.1), was conducted to confirm the latent factors of opportunity and propensity. The five items forming teacher-initiated activities for learning basic mathematics skills included teacher-reported frequency of counting out loud, using geometric manipulative, using counting manipulative, doing calendar activities, and discussing shapes and patterns. The five items forming teacher-initiated, integrated learning activities included teacher-reported frequency of doing mathematics games, music with math concept, creative movement with math, rules, cups and spoons, and telling time. A path model was used to test the effects of antecedent, opportunity, and propensity on prekindergarten mathematics achievement (see Fig. 12.1). The final set of parameter estimates and standard errors were obtained by combining the five sets of estimates using arithmetic rules outlined by Rubin (1987).

Second, in order to validate the multi-dimensional latent constructs of antecedent, opportunity, and propensity factors, first-order confirmatory factor analysis

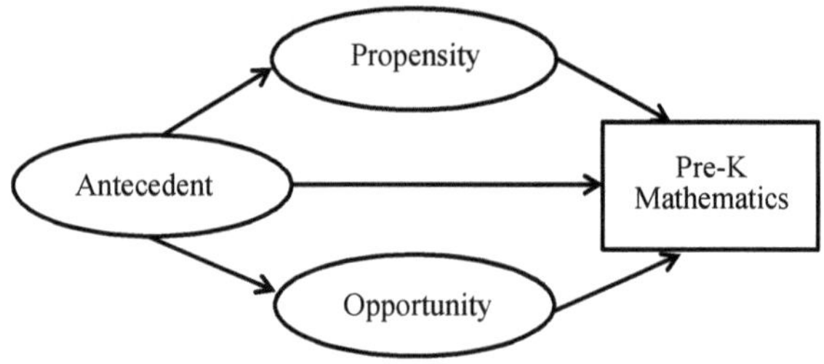

Fig. 12.1 Path model. Adapted from Wang, Shen and Byrnes (2013), with permission from Elsevier

(CFA) was conducted for the measurement model of antecedent, and second-order CFA was conducted for the measurement models of opportunity and propensity. The evaluation of the factor loadings showed that the observed indicators had high factor loadings on their common factors, indicating that they adequately reflected their underlying latent variables. All indicators in the model had statistically significant factor loadings (see Fig. 12.2), confirming the existence of significant associations among measured indicators and their latent constructs.

Third, in order to test if antecedent, opportunity, and propensity were important educational constructs, structural equation modeling (SEM) was performed. In our case, we examined whether the hypothesized O-P structural model provided a reasonable fit to the data by examining the goodness of fit between the hypothesized model and our data (Wang et al., 2013). A number of goodness-of-fit statistics were used, including the chi-square statistic or the Likelihood Ratio Test which examined the closeness of fit between the unrestricted sample covariance matrix and the restricted covariance matrix; the Comparative Fit Index (CFI) which evaluated the gain in improved fit between the null model and alternative model; and the Root Mean Square Error of Approximation (RMSEA) and its confidence interval which examined how well the hypothesized model would fit the population covariance matrix if it was available (Kline, 2011). Our results indicated a good fit. In everyday terms, the model you are trying to fit (see Fig. 12.1 for example) is evaluated by seeing if the relationships described in the model (the arrows pointing from one variable represented by an oval to another variable) seem to be present in the data itself. If there is a good fit, the theory or model is confirmed.

Fourth, we evaluated the direct effects, and as our fifth verification process, we evaluated the indirect and total effects of our model by evaluating the unstandardized and standardized validity coefficients for evaluation of sign, statistical and substantive significance.

Our last step was to evaluate the unique validity variance. In a prior study, Byrnes and Wasik (2009) found that their set of opportunity, propensity, and antecedent factors explained 63–71% of the variance of the outcome variable for a subsample of 17,401 US children (51% were female, 57% were White, 17% were Hispanic, 14% were Black, 6% Asian, and 6% were other) from ELCSK who had data on entry to kindergarten (age 5–6), end of kindergarten, grade one (age 6–7), and grade three (age 8–9). Stronger models explain more variance of the outcome variable. If the set of predictors do not really explain why some children learned

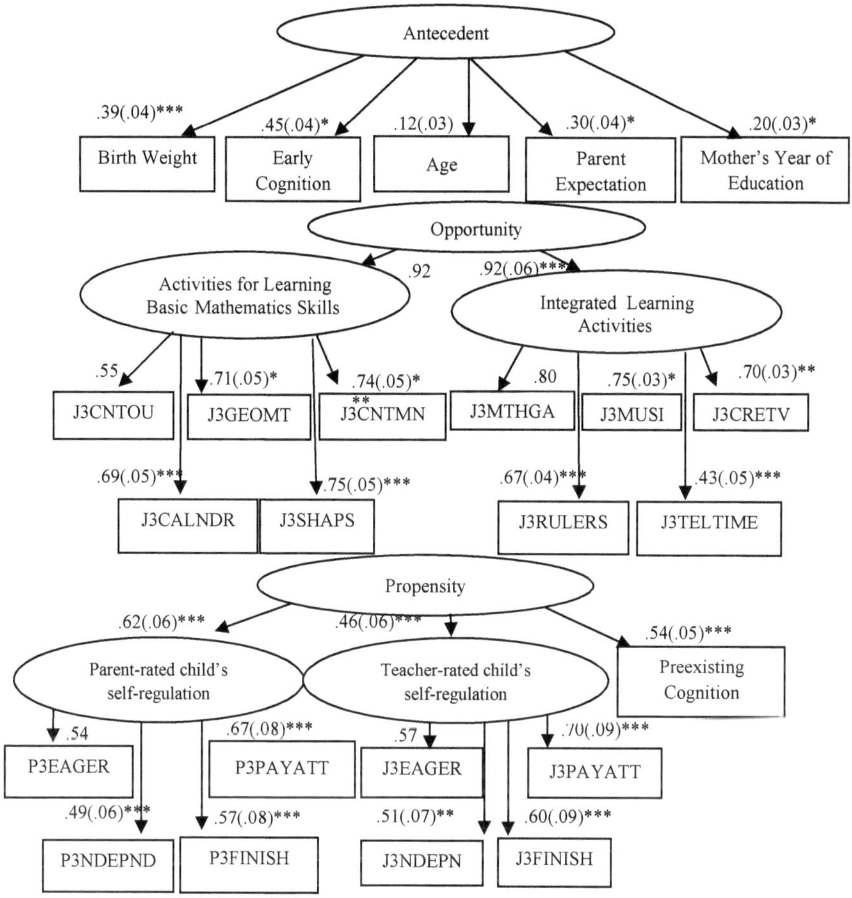

Fig. 12.2 Standardized factor loadings in the measurement models. Standard errors are reported in parentheses. ** $p < 0.01$; *** $p < 0.001$. Adapted Wang, Shen and Byrnes (2013), with permission from Elsevier

more than other children, 0% of the variance is explained. If all of the correct predictors are included, 100% of the variance is explained. So the O-P model seems to be gathering many of the correct, authentic predictors by explaining 63–71% of the variance in end-of-year mathematics scores. More specifically, at kindergarten (age 5–6), the propensity variables explained 69% of variance, opportunity variables explained less than 1% of variance, and antecedent explained less than 1% of variance. At grade one (age 6–7), propensity variables explained 46% of variance, opportunity explained less than 1% variance, and antecedent explained 1% of the variance. At grade three (age 8–9), propensity accounted for 46% variance, opportunity accounted for less than 1% variance, and antecedent accounted for 2% of variance.

In the Wang et al. (2013) study, the set of opportunity, propensity, and antecedent factors examined explained 34% of the variance. More specifically, the antecedent variables accounted for 21% of the variance; antecedent and propensity variables accounted for 31% of the variance; and antecedent, opportunity, and propensity variables accounted for 34% of the variance. The antecedent factor was positively associated with the opportunity and propensity factors. The opportunity factor was positively related to prekindergarten mathematics achievement. The total effect of the antecedent factor on prekindergarten mathematics achievement was also significant.

12.9 Testing and Extending the O-P Model on International Populations of Young Children

One particularly promising way to strengthen the generalizability of the O-P model is to test and extend the model on other populations of young children, using comparable multivariate longitudinal studies conducted by our international colleagues. While these researchers have yet to test and confirm the O-P model, their work can provide clues as to potential additional variables that we have not tested to date, and also see if there are variations in the predictive value of certain variables across cultures. Such variations can provide potentially useful information for theory revision and generalization. In our examination of the work of our international colleagues, we found our colleagues used different and promising authentic predictors for their opportunity, propensity, and antecedent variables (Table 12.1).

We highlight these promising authentic predictors below and invite our international colleagues to replicate and extend the OP model using their existing data sets from their context. Together, we can advance the field by generating evidence of convergent and divergent validity for the different factors of the theory, improving our measurement of these factors, strengthening our predictions, guiding the development of effective interventions, and supporting the successful application of theory to practice.

12.9.1 Opportunity Indicators Examined by International Colleagues

12.9.1.1 Home Numeracy Environment

Manolitsis et al. (2013) tested whether formal home numeracy environment will predict mathematics fluency in grade one on a sample of 82 Greek children (53 males and 29 females, mean age = 64.67 months, SD = 3.26, at the first time of measurement) from Heraklion, a typical urban city in Greece. The children were

Table 12.1 Promising authentic predictors examined by international colleagues

Study name	Population	Antecedent	Opportunity	Propensity
Alloway and Passolunghi (2011)	206 typically developing Italian children aged 7 and 8 (109 boys) recruited from four mainstream schools located in the northwest of Italy. The majority of parents came from professional homes that were predominantly middle class but included families from across the social spectrum. For the statistical analyses, participants were divided into two age-groups: 7-year-olds (n = 100; M = 88 months, SD = 3.5 months, 50 boys) and 8-year-olds (n = 106; M = 103 months, SD = 3.6 months, 46 boys). None was receiving special education services or had documented brain injury, or behavioral problems. None of the assessed children belonged to families with sociocultural disadvantage	Age	Not measured	General ability as measured by the vocabulary subtest of PMA (Thurstone & Thurstone, 1968). Working memory as assessed by 12 tests from the Automated Working Memory Assessment (AWMA, Alloway, 2007)
Aunio, Heiskari, Van Luit, and Vuorio (2015)	235 ordinary Finnish children (111 girls and 124 boys) around 6-years-old on average (in months, M = 74.55;	Not measured	Typical instruction offered during kindergarten from September until April	Prior numeracy skills, measured in September and January during kindergarten

(continued)

Table 12.1 (continued)

Study name	Population	Antecedent	Opportunity	Propensity
	SD = 3.50) who were a subsample of the Finnish Early Numeracy Test (ENT). The children in this study attended 18 ordinary Finnish kindergartens in the capital region. Children with a multilingual background (i.e., immigrant children) or those with special educational needs were excluded from the sample. The current data were collected at three measurement times (September, January, and April 2003) during 1 kindergarten year			
Carmichael, MacDonald, and McFarland-Piazza (2014)	2450 Australian kindergarten children (aged between 8.25 and 10; 53% male; 2% indigenous status) from the third wave of Longitudinal Study of Australian Children (LSAC) who completed their numeracy component of the Australian National Assessment Program—Literacy and Numeracy (NAPLAN) test for the first time	Parent education, family SES, home environment, parental warmth, parental hostility	Teacher qualification (education, years of teaching the grade), teacher self-reported math learning environment, children's attendance, Student–Teacher Relationship Scale, parent provision of academic support at home, parent and school interaction, parent–teacher interaction	IQ measured by the Matrix Reasoning Test, a component of the Wechsler Intelligence Scale for Children (WISC–IV) (Wechsler, 2003); child attitude toward school and child attitude toward math and number work at school

(continued)

Table 12.1 (continued)

Study name	Population	Antecedent	Opportunity	Propensity
Fitzpatrick and Pagani (2012)	1824 children (approx. 50% females with average age of 29 months) from the Québec Longitudinal Study of Child Development (QLSCD) with working memory data at 29 and 41 months	Child age, temperament, sleep, weight-for-gestational age, months breastfed, family functioning and configuration, and maternal age were entered and removed as did not significantly contribute to the variance in the model. Socioeconomic status and sex were retained	Not measured	Working memory as measured by the Imitation Sorting Task (Alp, 1994) assessed at 29 and 41 months by trained examiners
Manolitsis, Georgiou, and Tziraki (2013)	82 Greek children (53 males and 29 females; mean age = 64.67 months, SD = 3.26, at the first time of measurement) from Heraklion, a typical urban city in Greece, were followed from kindergarten to grade 1. The children were randomly selected from six kindergarten schools (serving children aged 5–6), which were, in turn, selected with a stratified randomized approach in order to represent a range of demographics. The children were native speakers of Greek, Caucasian, and 57% had attended prekindergarten (serving children aged 4–5)	Mother's education	Home numeracy environment (adapted in Greek from LeFevre et al. (2009) where parent self-report frequency of teaching child to identify numbers, count objects, sort objects, count, simple calculations)	Basic math concept (assessed by four tasks adapted from the Test of Early Mathematics Ability (TEMA-3; Ginsburg & Baroody, 2003). Verbal counting (child asked to count from 1 up to 50, or the higher number they could)

(continued)

Table 12.1 (continued)

Study name	Population	Antecedent	Opportunity	Propensity
Praet, and Desoete (2014)	132 Dutch-speaking children (53% girls) from five kindergartens serving children from families with working and middle-class-socioeconomic backgrounds	Not measured	Not measured	IQ measured by Wechsler Preschool and Primary Scale of Intelligence or the WPPSIIII-NL. Oral language skills measured by the Clinical Evaluation of Language Fundamentals or the CELF-4Nl (Kort et al., 2008; Semel et al., 2008). Estimations using number words and mathematics (in three formats, Arabic digits, dots, and number)
Wu and Chiang (2015)	19,499 children (53% were boys, 8% were preterm) who completed the 6-month, 18-month, and 3-year surveys of the Taiwan Birth Cohort Study (TBCS). The mean age of the mothers when giving birth was 28.88 years (SD = 4.35), and the majority had received either 10–12 years (40%) or more than 13 years (46%) of formal education	Family transition type (married stable, cohabiting stable, single stable, married unstable, single unstable); exposure to income poverty; family APGAR score; parenting quality (cognitive stimulation, emotional support	Not measured	Not measured

randomly selected from six kindergarten schools (serving children aged 5–6), which were, in turn, selected with a stratified randomized approach in order to represent a range of demographics. The children were native speakers of Greek, Caucasian, and 57% had attended prekindergarten (serving children aged 4–5). They measured formal home numeracy environment by adapting into Greek, the LeFevre et al. (2009) scale that measured parent self-report on frequency of teaching child to identify numbers, count objects, sort objects, count, and simple calculations. The results from their path analyses indicated parents' teaching of numeracy skills predicted mathematics fluency through the effects of verbal counting and that their overall model accounted for 26–27% of the variance (Manolitsis et al., 2013).

12.9.1.2 Supportive Home–School Relationship

Carmichael et al., (2014), in their exploratory study, measured how supportive home–school relationship, one of the factors in their exploratory ecological theory, would predict children's performance on a standardized numeracy assessment. They conducted their study on a subsample of 2,450 Australian kindergarten children (aged between 8.25 and 10; 53% male; 2% indigenous status) from the third wave of the Longitudinal Study of Australian Children (LSAC) who completed their numeracy component of the Australian National Assessment Program —Literacy and Numeracy (NAPLAN) test for the first time. They measured supportive home–school relationship through these measures: parent provision of academic support based on parent reports on two survey items, parent/school interaction based on teacher response to one survey item, and parent/teacher communication based on parents response to six survey items adapted from a measure in the Early Childhood Study of Kindergarteners (ECLS-K)—Base Year. They found parental involvement as measured by teachers' perception of parental involvement, parent help with homework, and parental communication explained 11% of the variation.

12.9.2 Propensity Indicators Examined by International Colleagues

We examined the literature to identify longitudinal studies of early mathematics achievement that were conducted by researchers outside of the USA and which focused on variables that have not been examined by O-P theorists based in the USA to date. Within disciplines such as educational psychology, mathematics education, and education policy, researchers adopt common definitions of theoretical constructs (e.g., "self-regulation," "inquiry learning," and "educational opportunities") and major journals in these disciplines publish the work of scholars from around the world. As such, the studies described below refer to these shared constructs. That said, the relative weight of the shared constructs could differ by country (e.g., family income could be a stronger predictor of performance in countries that have wider income disparities—as has been found in PISA); these constructs could be manifested, understood, and measured in distinct ways in these countries; there may be factors in each category that are unique to particular countries that await to be identified. Our brief review is intended to initiate the dialogue and welcome collaborations that help reveal these national differences to help extend and improve the model.

12.9.2.1 Working Memory

Alloway and Passolunghi (2011) extended current understanding of the relationship between working memory and mathematical skills in their investigation of the contribution of working memory and verbal ability to mathematical skills on a sample of 206 typically developing Italian children aged 7 and 8 (109 boys) recruited from four mainstream schools located in the northwest of Italy. Working memory was assessed by 12 tests from the Automated Working Memory Assessment (AWMA, Alloway, 2007), a computer-based standardized battery that provides three assessments of verbal short-term memory, three assessments of visuospatial memory, three assessments of verbal working memory, and three assessment of visuospatial working memory. They found that working memory accounted for four to nine percent of the variance and that the general model accounted for 17–35% of the variance (Alloway & Passolunghi, 2011).

Fitzpatrick and Pagani (2012) tested their hypothesis that better working memory skills at 35 months will predict better kindergarten classroom engagement and academic performance on a subsample of 1824 children (approx. 50% females with average age of 29 months) from the Québec Longitudinal Study of Child Development (QLSCD). Working memory was measured by the Imitation Sorting Task (Alp, 1994) and was assessed at 29 and 41 months by trained examiners. Using stepwise regression, they found their model explained 17–35% of the variance of mathematics skills with general ability explaining around 13–26% of the variance and working memory explaining around 4–9% of the variance. Specifically, they found a positive association between early working memory scores and later classroom engagement, number knowledge, and receptive vocabulary.

12.9.2.2 Quality of Parent–Child Relationship

Wu and Chiang (2015) examined the relationships and potential pathways between family structure transitions and early childhood development on the 19,499 children (53% were boys, 8% were preterm births) who completed the 6-month, 18-month, and 3-year surveys of the Taiwan Birth Cohort Study (TBCS). The mean age of the mothers when giving birth was 28.88 years (SD = 4.85), and the majority had received either 10–12 years (40%) or more than 13 years (46%) of formal education. They measured family transition type (married stable, cohabiting stable, single stable, married unstable, and single unstable) and parenting quality (degree of cognitive stimulation and emotional support). They found caregiver's psychosocial functioning and/or parenting quality can partly buffer against the risks of poorer cognitive or socioemotional development that accompany certain family transition types.

12.10 Conclusion and Implications for Theory Building, Policy and Practice

Earlier we noted that an important first step in knowing how to increase the mathematics skills of young children and thereby equip them for later success in school is to build a theory using the results of large-scale, longitudinal multivariate studies. The process taken so far is to add variables to the model that explain additional variance once other consistent predictors have been included and controlled, and delete variables that drop out when other more powerful and authentic predictors have been included (e.g., race after income is controlled). At present, the set of approximately 20 predictors in the three categories that remain significant account for about 50–60% of the variance in mathematics achievement (e.g., Byrnes & Miller-Cotto, 2016; Wang et al., 2013). Hence, much is known about why individuals and groups differ in achievement, but additional variables wait to be discovered and their relative importance in each country needs to be established. Once these additional variables are discovered and the amount of explained variance moves closer to 100%, scholars can then propose and test a theoretical model that effectively integrates the variables and describes how they relate to each other. This simple yet versatile model can then be tested in experimental investigations to verify the causal role of each of the variables. Once scholars and practitioners know why certain children show higher levels of mathematics achievement than others by the end of the preschool period, effective forms of intervention can be created to target the causal factors identified in the theory.

In prior studies of the O-P framework, we identified and verified the predictive role of antecedent factors such as family socioeconomic status (family income and maternal education), parent educational expectations for their children, age, birth weight, gender, and ethnicity. With respect to opportunity factors, we identified and confirmed the predictive role of several aspects of instruction (the content that is presented, the style of presentation, and the dosage per week of either a full curriculum or supplemental activities). Our international colleagues have identified several others as well as the ones we describe in this chapter: home numeracy environment and supportive home–school relationships. With respect to propensity factors, we have identified and confirmed the role of prior knowledge, motivation, and self-regulation in our studies.

We encourage other researchers to build on this work in order to create the most accurate, predictive, and generalizable model of early children mathematics achievement. This way, we can collaborate in guiding early mathematics policy-makers, practitioners, and professional and advocacy organizations by providing them with a simple yet flexible framework on early mathematics achievement to scaffold understanding, generate and test hypotheses, and facilitate and adapt targeted interventions to address context- and cultural-specific problems.

References

Alexander, K. L., Entwisle, D. R., & Dauber, S. L. (1993). First-grade classroom behavior: Its short- and long-term consequences for school performance. *Child Development, 64*(3), 801–814.

Alloway, T. P. (2007). *Automated working memory assessment*. London: Harcourt Assessment.

Alloway, T. P., & Passolunghi, M. C. (2011). The relationship between working memory, IQ, and mathematical skills in children. *Learning and Individual Differences, 21,* 133–137. doi:10.1016/j.lindif.2010.09.013.

Alp, I. E. (1994). Measuring the size of working memory in very young children: The imitation sorting task. *International Journal of Behavioral Development, 17,* 125–141.

Aunio, P., Heiskari, P., Van Luit, J. E. H., & Vuorio, J. M. (2015). The development of early numeracy skills in kindergarten in low-, average- and high-performance groups. *Journal of Early Childhood Research, 13*(1), 3–16. doi:10.1177/1476718X14538722.

Bollen, K. A. (2011). Evaluating effect, composite, and causal indicators in structural equation models. *MIS Quarterly, 35*(2), 359–372.

Bowen, J., Gibson, F., & Hand, P. (2002). Educational outcome at 8 years for children who were born extremely prematurely: A controlled study. *Journal of Paediatrics and Child Health, 38* (5), 438–444.

Byrne, B. M. (1998). *Structural equation modeling with LISREL, PRELIS, and SIMPLIS: Basic concepts, applications, and programming*. New York, NY: Lawrence Erlbaum Associates, Taylor & Francis Group.

Byrnes, J. P. (2003). Factors predictive of mathematics achievement in White, Black, and Hispanic 12th graders. *Journal of Educational Psychology, 95,* 316–326.

Byrnes, J. P. (2011). Academic achievement. In B. Brown & M. Prinstein (Eds. in chief), *Encyclopedia of Adolescence* (pp. 1–9). Amsterdam: Elsevier.

Byrnes, J. P., & Miller, D. C. (2007). The relative importance of predictors of mathematics and science achievement: An opportunity-propensity analysis. *Contemporary Educational Psychology, 32,* 599–629. doi:10.1016/j.cedpsych.2006.09.002.

Byrnes, J. P., & Miller Cotto, D. (2016). The growth of mathematics and reading skills in segregated and diverse schools: An opportunity-propensity analysis of national database. *Contemporary Educational Psychology, 46,* 34–51. doi:10.1016/j.cedpsych.2016.04.002

Byrnes, J. P., & Wasik, B. A. (2009). Factors predictive of mathematics achievement in kindergarten, first and third grades: An opportunity-propensity analysis. *Contemporary Educational Psychology, 34,* 167–183. doi:10.1016/j.cedpsych.2009.01.002.

Carmichael, C., MacDonald, A., & McFarland-Piazza, L. (2014). Predictors of numeracy performance in national testing programs: Insights from the longitudinal study of Australian children. *British Educational Research Journal, 40*(4), 637–659. doi:10.1002/berj.3104.

Catsambis, S. (2001). Expanding knowledge of parental involvement in children's secondary education: Connections with high school seniors' academic success. *Social Psychology of Education, 5*(2), 149–177.

Chard, D. J., Baker, S. K., Clarke, B., Jungjohann, K., Davis, K., & Smolkowski, K. (2008). Preventing early mathematics difficulties: The feasibility of a rigorous kindergarten mathematics curriculum. *Learning Disability Quarterly, 31*(1), 11–20 (Mathematics Instruction and Learning Disabilities).

Clarke, B., Smolkowski, K., Baker, S., Fien, H., Doabler, C., & Chard, D. (2011). The impact of a comprehensive tier I core kindergarten program on the achievement of students at risk in mathematics. *The Elementary School Journal, 111*(4), 561–584.

Clements, D. H., & Sarama, J. (2007). Effects of a preschool mathematics curriculum: Summative research on the Building Blocks Project. *Journal for Research in Mathematics Education, 38* (2), 136–163.

Clements, D. H., & Sarama, J. (2008). Experimental evaluation of the effects of a research-based preschool mathematics curriculum. *American Educational Research Journal, 45*(2), 443–494.

Clements, D. H., & Sarama, J. (2009). *Learning and teaching early math: The learning trajectories approach*. New York: Routledge.

Clements, D. H., Sarama, J., Spitler, M. E., Lange, A. A., & Wolfe, C. B. (2011). Mathematics learned by young children in an intervention based on learning trajectories: A large-scale cluster randomized trial. *Journal for Research in Mathematics Education, 42*(2), 127–166.

Coleman, J. S. (1987). Families and schools. *Educational Researcher, 16,* 32–38.

Collins, J., & David, R. (1990). The differential effect of traditional risk factors on infant birth weight among Blacks and Whites in Chicago. *American Journal of Public Health, 80,* 679–681.

Davis-Kean, P. E. (2005). The influence of parent education and family income on child achievement: The indirect role of parental expectations and the home environment. *Journal of Family Psychology, 19*(2), 294–304.

DiPerna, J. C., Volpe, R. J., & Elliott, S. N. (2005). A model of academic enablers and mathematics achievement in the elementary grades. *Journal of School Psychology, 43,* 379–392.

Dowker, A. (2008). Individual differences in numerical abilities in preschoolers. *Developmental Science, 11*(5), 650–654.

Duncan, G. J., Dowsett, C. J., Claessens, A., Magnuson, K., Huston, A. C., Klebanov, P., …, Japel, C. (2007). School readiness and later achievement. *Developmental Psychology, 43*(6), 1428–1446.

Duncan, G. J., Morris, P. A., & Rodrigues, C. (2011). Does money really matter? Estimating impacts of family income on young children's achievement with data from random-assignment experiments. *Developmental Psychology, 47*(5), 1263–1279. doi:10.1037/a0023875.

Early, D., Barbarin, O., Bryant, D., Burchinal, M., Chang, F., Clifford, R., … Barnett, W. S. (2005). *Pre-kindergarten in seven states: NCEDL's multi-state study of pre-kindergarten & study of state-wide early education programs (SWEEP)*. Chapel Hill: NC: University of North Carolina.

Eccles, J. S. (2005). Influences of parents' education on their children's educational attainments: The role of parent and child perceptions. *London Review of Education, 3*(3), 191–204. doi:10.1080/14748460500372309.

Farran, D. C., Lipsey, M. W., Watson, B., & Hurley, S. (2007). *Balance of content emphasis and child content engagement in an early reading first program*. Paper presented at the American Educational Research Association.

Fehrmann, P. G., Keith, T. Z., & Reimers, T. M. (1987). Home influence on school learning: Direct and indirect effects of parental involvement on high school grades. *Journal of Educational Research, 80*(6), 330–337.

Fitzpatrick, C., & Pagani, L. S. (2012). Toddler working memory skills predict kindergarten school readiness. *Intelligence, 40,* 205–212.

Furstenberg, F. F., Cook, T. D., Eccles, J., Elder, G. H., & Sameroff, A. (1999). *Managing to make it: Urban families and adolescent success*. Chicago, IL: University of Chicago.

Geary, D. C. (2007). An evolutionary perspective on learning disability in mathematics. *Developmental Neuropsychology, 32*(1), 471–519.

Georges, A. (2009). Relation of instruction and poverty to mathematics achievement gains during kindergarten. *Teachers College Record, 111*(9), 1–12.
Graham, T. A., Nash, C., & Paul, K. (1997). Young children's exposure to mathematics: The child care context. *Early Childhood Education Journal, 25,* 31–38.
Hailikari, T., Nevgi, A., & WKomulainen, E. (2008). Academic self-beliefs and prior knowledge as predictors of student achievement in mathematics: A structural model. *Educational Psychology, 28*(1), 59–71.
Hillemeier, M., Morgan, P., Farkas, G., & Maczuga, S. (2010). Perinatal and socioeconomic risk factors for variable and persistent cognitive delay at 24 and 48 months of age in a national sample. *Maternal and Child Health Journal,* 1–10.
Hindman, A. H., Skibbe, L. E., Miller, A., & Zimmerman, M. (2010). Ecological contexts and early learning: Contributions of child, family, and classroom factors during Head Start, to literacy and mathematics growth through first grade. *Early Childhood Research Quarterly, 25,* 235–250.
Jones, K. K., & Byrnes, J. P. (2006). Characteristics of students who benefit from high-quality mathematics instruction. *Contemporary Educational Psychology, 31*(3), 328–343.
Jordan, N. C., Kaplan, D., Olah, L. N., & Locuniak, M. N. (2006). Number sense growth in kindergarten: A longitudinal investigation of children at risk for mathematics difficulties. *Child Development, 77*(1), 153–175. doi:10.1111/j.1467-8624.2006.00862.x.
Kaestle, C. F. (1993). The awful reputation of education research. *Educational Researcher, 22*(1), 26–31.
Kline, R. B. (2011). Principles and practice of structural equation modeling (3rd ed.). New York, NY: The Guilford Press.
Kort, W., Schittekatte, M., & Compaan, E. (2008). CELF-4-NL Test voor diagnose en evaluatie van taalproblemen. Handleiding [Test for the evaluation of language problems. Manual]. Amsterdam: Pearson.
Kurdek, L. A., & Sinclair, R. J. (1988). Relation of eighth graders' family structure, gender, and family environment with academic performance and school behavior. *Journal of Educational Psychology, 80*(1), 90–94.
Lamy, C. E., Frede, E., Seplocha, H., Strasser, J., Jambunathan, S., Juncker, J. A., et al. (2004). *Inch by inch, row by row, gonna make this garden grow: Classroom quality and language skills in the Abbott Preschool Program.* Trenton, NJ: New Jersey Department of Education, Office of Early Childhood Education.
Lee, V. E., & Burkam, D. T. (2002). *Inequality at the starting gate: Social background differences in achievement as children begin school.* Washington, DC: Economic Policy Institute.
LeFevre, J., Skwarchuk, S., Smith-Chant, B., Fast, L., Kamawar, D., & Bisanz, J. (2009). Home numeracy experiences and children's math performance in the early school years. *Canadian Journal of Behavioural Science, 41,* 55–66.
Lung, F-W., Shu, B-C, Chiang, T-L., Y Lin, S-J. (2011). Maternal mental health and childrearing context in the development of children at 6, 18 and 36 months: A Taiwan birth cohort pilot study. *Child: Care, Health and Development, 37*(2), 211–223. doi:10.1111/j.1365-2214.2010.01163.
Manolitsis, G., Georgiou, G. K., & Tziraki, N. (2013). Examining the effects of home literacy and numeracy environment on early reading and math acquisition. *Early Childhood Research Quarterly, 28,* 692–703. doi:10.1016/j.ecresq.2013.05.004.
McClelland, M. M., Cameron, C. E., Connor, C. M., Farris, C. L., Jewkes, A. M., & Morrison, F. J. (2007). Links between behavioral regulation and preschoolers' literacy, vocabulary, and mathematics skills. *Developmental Psychology, 43,* 947–959.
Murphy, P. K., Alexander, P. A. (2002). What counts? The predictive powers of subject-matter knowledge, strategic processing, and interest in domain-specific performance. *Journal of Experimental Education, 70*(3), 197–214.
National Association for the Education of Young Children. (2002). *Early childhood mathematics: Promoting good beginnings.* Washington, D. C.: Author.

Perry, M. A., & Fantuzzo, J. W. (2010). A multivariate investigation of maternal risks and their relationship to low-income, preschool children's competencies. *Applied Developmental Science, 14*(1), 1–17.

Praet, M., & Desoete, A. (2014). Number line estimation from kindergarten to grade 2: A longitudinal study. *Learning and Instruction, 33,* 19–28. doi:10.1016/j.learninstruc.2014.02.003.

Ramani, G. B., & Siegler, R. S. (2008). Promoting broad and stable improvements in low-income children's numerical knowledge through playing number board games. *Child Development, 79* (2), 375–394.

Ramani, G. B., & Siegler, R. S. (2011). Reducing the gap in numerical knowledge between low- and middle-income preschoolers. *Journal of Applied Developmental Psychology, 32*(3), 146–159.

Ransdell, S., & Hecht, S. (2003). Time and resource limits on working memory: Cross-age consistency in counting span performance. *Journal of Experimental Child Psychology, 86*(4), 303–314. doi:10.1016/j.jecp.2003.08.002.

Rubin, D. B. (1987). *Multiple imputation for nonresponse in surveys*. New York: Wiley.

Rutchick, A. M., Smyth, J. M., Lopoo, L. M., & Dusek, J. B. (2009). Great expectations: The biasing effects of reported child behavior problems on educational expectancies and subsequent academic achievement. *Journal of Social and Clinical Psychology, 28*(3), 392–413.

Sajaniemi, N., Hakamies-Blomqvist, L., Mäkelä, J., Avellan, A., Rita, H., & von Wendt, L. (2001). Cognitive development, temperament and behavior at 2 years as indicative of language development at 4 years in pre-term infants. *Child Psychiatry and Human Development, 31*(4), 329–346.

Schunk, D. H., & Zimmerman, B. J. (2013). Self-regulation and learning. In W. M. Reynolds, G. E. Miller, & I. B. Weiner (Eds.), *Handbook of psychology, Vol. 7: Educational psychology* (2nd ed.) (pp. 45–68). New York, NY: Wiley.

Semel, E., Wiig, E. H., & Secord, W. A. (2008). CELF 4 Nl clinical evaluation of language fundamentals. Amsterdam: Pearson.

Shavelson, R. J., & Towne, L. (2002). *Scientific research in education*. Washington, D. C.: National Academy Press.

Siegler, R. S., & Ramani, G. B. (2009). Playing linear number board games—but not circular ones—improves low-income preschoolers' numerical understanding. *Journal of Educational Psychology, 101,* 545–560.

Soares, D. L., Lemos, G. C., Primi, R., & Almeida, L. S. (2015). The relationship between intelligence and academic achievement throughout middle school: The role of students' prior academic performance. *Learning and Individual Differences, 41,* 73–78.

Starkey, P., Klein, A., & Wakeley, A. (2004). Enhancing young children's mathematical knowledge through a pre-kindergarten mathematics intervention. *Early Childhood Research Quarterly, 19*(1), 99–120.

Sternberg, R. J., Grigorenko, E., & Bundy, D. A. (2001). The predictive value of IQ. *Merrill-Palmer Quarterly,* 1–41.

Thurstone, N. L., & Thurstone, T. G. (1968). PMA Primary Mental ability (trad. it. Batteria Primaria di Abilità). Firenze: Organizzazioni Speciali.

Tudge, J. R. H., & Doucet, F. (2004). Early mathematical experiences: Observing young Black and White children's everyday activities. *Early Childhood Research Quarterly, 19,* 21–39.

Walters, P. B., & Lareau, A. (2009). Introduction. In P. B. Walters, A. Lareau, & S. H. Ranis (2009), *Educational research on trial: policy reform and the call for scientific rigor* (pp. 1–13). New York, NY: Routledge.

Wang, A. H. (2010). Optimizing early mathematics experiences for children from low-income families: A study on opportunity to learn. *Early Childhood Education Journal, 37*(4), 295–302. doi:10.1007/s10643-009-0353-9.

Wang, A. H, Firmender, J. M., Power, J. R., & Byrnes, J. P. (2016). Understanding the characteristics and effects of school-based early mathematics learning opportunities in early childhood settings: A meta-analytic review. *Early Education and Development*, doi:10.1080/10409289.2016.1116343.

Wang, A. H., Shen, F., & Byrnes, J. (2013). Does the opportunity-propensity framework predict early mathematics skills for low-income pre-kindergarten children? *Contemporary Educational Psychology, 38,* 259–270. doi:10.1016/j.cedpsych.2013.04.004.

Wechsler, D. (2003). Wechsler intelligence scale for children: fourth edition (San Antonio, TX: Psycho-logical Corporation).

Wigfield, A., & Cambria, J. (2010). Students' achievement values, goal orientations, and interest: Definitions, development, and relations to achievement outcomes. *Developmental Review, 30* (1), 1–35.

Wood, D., Kurtz-Costes, B., & Copping, K. E. (2011). Gender differences in motivational pathways to college for middle class African American youths. *Developmental Psychology, 47* (4), 961–968.

Wu, J. C., & Chiang, T. L. (2015). Family structure transitions and early childhood development in Taiwan: Evidence from a population-based birth cohort study. *International Journal of Behavioral Development, 39*(3), 275–284. doi:10.1177/0165025414544.

Authors Biography

Dr. Aubrey H. Wang is Associate Professor and Director of Doctor of Education Program at Saint Joseph's University. Aubrey has worked in educational policy and research, and educational leadership for 20 years. Aubrey has worked as a research team member at the Consortium for Research in Education and the Educational Testing Service. Aubrey's research interests include how to use research and policy to improve early mathematics learning, school readiness, and achievement gap. Currently, she is leading an international team to investigate causes of school success and discrepancies for young children in the USA, Canada, and Taiwan using multivariate analyses.

Dr. James P. Byrnes is Professor of Educational Psychology and Applied Developmental Science at Temple University in Philadelphia, PA, USA. His research focus has been on developing and testing the Opportunity-Propensity Model of achievement in a variety of small-scale and large-scale multivariate studies. Other areas of interest include the importance of conceptual knowledge in acquiring mathematical skills, the information value of neuroscientific findings for psychological theories, and developing a model to explain adolescent and adult risk-taking. He is the proud spouse of Barbara Wasik, an expert in early childhood development, and the proud parent of Tom and Julia.

Chapter 13
Funds of Knowledge: Children's Cultural Ways of Knowing Mathematics

Maulfry Worthington

Abstract In the prevailing global climate, many teachers feel pressured to demonstrate children's 'basic skills' that often result in direct teaching and a marginalization of play, and widespread orthodoxy means that direct teaching of mathematical notations continues, often causing children considerable problems. Yet research has shown that the beginnings of the abstract symbolic language of mathematics have their roots in young children's home cultural experiences, extended in meaningful contexts of pretend-play and other child-initiated activities. This chapter draws on findings from recent doctoral research into the beginnings of young children's mathematical semiosis in their homes and nursery school, revealing the power of their mathematical thinking and understandings expressed through their graphical communications. The chapter focuses on the social and cultural contexts of home and pretend-play, the children's mathematical graphics underpinned by Vygotsky's cultural–historical theory, and his dialectical view of relationships between play and symbolic tool-use. Understandings of the abstract symbolic language of mathematics are social and cultural, this chapter arguing that children's personal mathematical communications evolve over time. Competency develops as a continuum, revealing how children's early understandings contribute to subsequent mathematical notations. This study chapter uncovers the beginnings of abstraction in early childhood mathematics, focusing on the importance of children's interests and their cultural knowledge in underpinning their subject knowledge. It argues that spontaneous, social pretend-play be better understood for supporting children's interests and for its mathematical potential, and for children's mathematics be prioritized in early childhood curricula so that their existing competencies and informal representations are valued and understood.

Keywords Cultural knowledge · Children's interests · Pretend-play
Preschool · Informal mathematics

M. Worthington (✉)
Vrije Universiteit, Amsterdam, The Netherlands
e-mail: maulfry@icloud.com

V. Kinnear et al. (eds.), *Forging Connections in Early Mathematics Teaching and Learning*, Early Mathematics Learning and Development,
https://doi.org/10.1007/978-981-10-7153-9_13

13.1 Introduction

Mathematical semiosis refers to the ways in which we use a range of ways to express meanings through, for example, gesture and language, children's graphical marks, symbols and signs. Since mathematics and writing are human products they are inseparable from their cultural context: we learn through participating in meaningful cultural practices (Rogoff, 2008). This suggests that social pretend-play might offer contexts for children to explore their mathematical 'funds of knowledge' (Moll et al., 1992); such play is strongly linked to children's interests (Hedges et al., 2011) and latterly to their cultural knowledge of mathematics (Worthington, 2015b; Worthington & van Oers, 2016). Vygotsky emphasized that writing (and by analogy, mathematics):

> … must be 'relevant for life', [it] should be meaningful for children, that an intrinsic need should be aroused in them, and that writing should be incorporated into a task that is necessary and relevant for life. In the same way as children learn to speak, they should be able to learn to read and write…. Reading and writing should be necessary for her in her play (1978, p. 118).

To what extent can pretend-play support children's *relevant*, *meaningful* and *necessary* mathematics? The following example serves as an illustration and as an introduction to young children's ability to deal with the abstractions that underpin mathematics, exemplifying themes explored throughout the chapter.

13.1.1 Play Narrative 1

Triggered by Isaac's interest in maps, Isaac and David created a large plan of a road layout (Fig. 13.1) and making lines Isaac announced, '*These are roads crossing other roads*'. Then number a square with a series of lines explained, '*These are arrows to say "go this way"*'. He drew another square on another part of the paper, '*This is the car park gate; the square on the outside means it's shut*'. The following day Isaac '*opened*' the gate by making the two horizontal lines [across the 'gate' he'd drawn]. Together they created imaginary accidents and problems with the vehicles such as crashes. David's helicopter rescued the people (little wooden play people), while Isaac stood them all up in a crowd '*to watch*'. Pointing, Isaac explained, '*here's where you can park your lorries for 2 h while you sit on the beach*'. Then David referred to the charge for the car park, *adding* '*We're going to have signs to say what the speed limit is*'.

This example is particularly interesting for its evidence of abstract thinking through the boys' *explorations with symbols*.[1] For example, Isaac 'opened' his gate

[1]Terms in italics that follow each figure refer to domains in Carruthers and Worthington's taxonomy (2013).

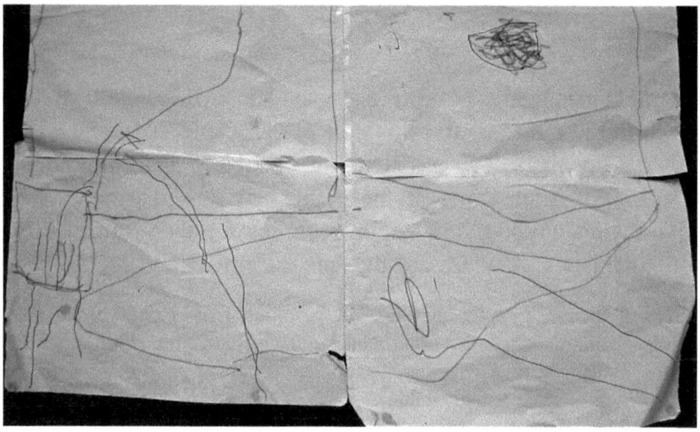

Fig. 13.1 Isaac and David's plan of a road layout, Carruthers and Worthington, *Understanding Children's Mathematical Graphics,* 2011. Reproduced with the kind permission of Open International Publishing Ltd.

by adapting the symbol he'd drawn, using spoken words to express that his symbols signified particular meanings: 'these are arrows *to say* "go this way"'; the square on the outside *means* it's shut'. This ability to invent and adapt symbols and signs stands the children in good stead when they later work on mathematical operations (Carruthers & Worthington, 2008). For example, Worthington and Carruthers refer to children's use of arrows as 'narrative actions' (2003). Children sometimes choose to use arrows to indicate the operation of adding or subtracting items (Hughes, 1986), some also using them in place of the symbols for 'add' or subtract', (Carruthers & Worthington, 2005, 2006). Poland et al. describe arrows as one of the numbers of 'dynamic schematizations', where children use them to signify transformation, movement or change: they argue that such symbols should be given emphasis in early childhood since they underpin 'most mathematical activities' (2009, p. 310).

Isaac and David's teacher Emma commented that they, '*...invent[ed] their own symbols...Their persistence and excitement continued through the week and again Isaac, interested in signs, thought of his own "opening" sign. They were involved in high-level play that they initiated and they wove imaginary stories and challenging problems that they solve. They were totally engrossed as they organized the plan carefully, using any materials they needed to further develop their play and narratives. Isaac has a sophisticated understanding of roads that has enabled him to map roads and pathways with graphic representations. He has a growing understanding of road requirements too, such as the speed limits and crossroads. He uses his understanding of different marks to represent different purposes... The culture of the setting encouraged them to return to their theme over several weeks. They were bringing everything they knew about beaches, car parks and helicopters to their play and were completely free to invent the symbols they needed for their play*'.

13.2 Background

Learning mathematics begins at birth (e.g. van Oers, 2013; Lakoff & Nunez, 2001). It develops in the home (e.g. Worthington & van Oers, 2016; Street et al., 2008; Aubrey, 1997; Carruthers, 1997), extending to society and employment (Lave & Wenger, 1991). Although some of the mathematics used by adults (particularly those who are un-schooled) may appear similar in their informality to young children's mathematics, they are not the same. Empson and Jacobs observe how,

> '… the kinds of strategies, representations and reasoning used by children often differ from those used by mathematicians and other adults… Thus, by saying children's mathematics, we imply the existence of a coherent and logical approach to reasoning that differs in important ways from that of mathematicians and other adults'. (2008, p. 260, emphasis in the original)

Following publication of Vygotsky's work in the west, interest in 'authentic' and meaningful contexts for young children's mathematics grew in the UK. (e.g. Clayden et al., 1994; Munn, 1994; Hughes, 1986). Research identified problems children experience with a 'transmission' approach, particularly when teaching involves symbolic languages such as writing and mathematics (e.g. Poland, 2007; Carruthers & Worthington, 2006; Hughes, 1986). Brooker argues instead for 'learning as a*cquisition* (child-led)', learning as 'the *transformation of participation in cultural activities*' (2010, p. 41, emphasis in the original).

In his seminal work, Hughes investigated the difficulties young children have with formal mathematical notations and language, demonstrating how they could use personal marks and symbols to represent small quantities, and later represent calculations in informal ways. He argued that 'these representations are often ingenious and of considerable personal significance to children [and] should be the basis of any early work on written symbolism' (1986, p. 177). Aubrey concurred, 'since the child's mathematical knowledge cannot be separated from the situations in which it was acquired, mathematics will need to be embedded in the context of known situations and authentic activities' (1993, p. 30). Munn and Kleinberg argue that children need to learn the cultural rules of mathematics concerning, 'how to use a system, and what its role is in our culture… These cultural rules are possibly the most important things that children learn': without understanding these rules children 'risk becoming stranded in a sea of meaningless activity' (2003, pp. 51/53).

Three educational approaches to meaningful mathematical notations are rooted in the cultural–historical research of Vygotsky and his followers, and all prioritize children's understandings of mathematical inscriptions, children beginning with their own symbols, signs and representations. These approaches include 'Realistic Mathematics Education' (RME) (van den Heuvel-Panhuizen et al., 2014); children's 'schematizing' in pretend-play (e.g. van Oers, 2012; Poland, 2007; Poland et al., 2009), and *children's mathematical graphics*, (Carruthers and Worthington, (e.g. 2005, 2006).

As for the two approaches above, *children's mathematical graphics* emphasizes meaningful contexts, children's mathematical thinking, communication and problem-solving, but diverge in their pedagogical approaches. In their *mathematical graphics*, young children begin with their own intuitive and informal representations, attaching mathematical meanings to them from two to three years of age: the learning environment and increasingly their teachers mediate children's understandings of their marks and notations.

During the course of their research, Carruthers and Worthington analysed over 700 examples of *children's mathematical graphics*, charting children's symbolic representations from birth to eight years from their earliest gestures, marks, symbols and signs through to calculations and problem-solving (Carruthers & Worthington, 2005, 2013; Worthington & Carruthers, 2003). This was the first time that children's developing sign use for mathematics had been charted. However, until now neither the early beginnings of young children's *mathematical graphics* nor their cultural foundations have been investigated.

13.2.1 Context of the Current Research

The research featured in this chapter draws on longitudinal, ethnographic data gathered from an inner city maintained nursery in the southwest of England. Data include written observations and graphics of seven children aged three to four years of age, their teachers' documentations supplemented by additional observations made by the author. These 'learning diaries' (Carr, 2001) form the main body of data.

Three of the children's graphics will be scrutinized in detail in this chapter; these three children are Ayaan from Somalia, Shereen from the Philippines and Isaac. Analysis is interpretive, following a social semiotic paradigm and also draws on Carruthers and Worthington's taxonomy (2013). Ethical guidelines have been adhered to throughout. Analysis of data from the children in the present study highlight specific features of their *mathematical graphics*, showing the relationship between the cultural literacy *practices* the children experienced at home and the social literacy *events* (Street et al., 2008) within their play narratives, in which the children spontaneously engaged: many include the children's use of *mathematical graphics* to communicate.

13.3 Subject Knowledge in Early Childhood

Hedges and Cullen (2005) observe that the idea of 'subjects' has long been controversial in early childhood education, polarizing beliefs of those who feel they are inappropriate for young learners and have no place in a holistic child-centred approach (e.g. Nutbrown, 1999; Curtis, 1998), while for others the outcomes of subject learning are seen as desirable (Marcon, 2002). Hedges and Cullen (2005) propose that play be integrated with other 'subject' knowledge while Fleer recommends a 'wholeness approach', built on Vygotsky's understanding (1998) of children's development as an 'integrated whole' (2010, p. ix), The 'whole' includes both the individual and the collective, which 'must be considered together', encompassing 'the complexity of children in their everyday lives, and how caregivers enter into a relationship with children to create children's developmental life course' (Fleer, 2010, p. viii). Christie and Roskos (2009) call for the integration of play and literacy in early education and Pramling and Johansson (2006) and other researchers share this view.

While children develop their mathematical 'funds of knowledge' at home (Worthington & van Oers, 2016), their teachers are often not aware of this rich cultural knowledge (Aubrey, 1997). Moyles, Adams and Musgrove (2002) found that managers/head teachers, practitioners and parents 'overwhelmingly' believed subject curriculum knowledge to be less important than knowledge and understanding of children. In contrast, findings from research by Hedges and Cullen showed that whereas there was widespread support for the importance of subject knowledge, a sociocultural stance could resolve the polarization of philosophical positions of 'child-initiated' versus 'academics… *particularly if the content relates to children's interests*' (2005, pp. 75/77, emphasis added).

13.4 Children's Interests

Early childhood researchers have emphasized the importance of *children's interests* (e.g. Bertram & Pascal, 2002; Tizard & Hughes, 1984), which Hedges et al. (2011) construe as 'children's spontaneous, self-motivated play, discussions, enquiry, and or investigations that derive from their social and cultural experiences' (p. 187). Early childhood curricula increasingly refer to *children's interests*, for example, in England, a recent government document refers to the importance of creating 'an ethos which respects each child as an individual and which *values children's efforts, interests and purposes as instrumental to successful learning*' (DfE, 2013, p. 9, emphasis added). In New Zealand, *Te Whāriki* also emphasizes children's interests (Ministry of Education, 1996), and in Sweden, activities 'should be based on the child's experiences, interests, needs and views' (National Agency for Education, 2011, p. 11).

Young children develop their interests at home and in the community through culturally valued and purposeful activities: for children these are often with a parent, and sometimes with grandparents and siblings. Lave and Wenger see such involvement as 'legitimate peripheral participation' (Lave and Wenger, 1991), with David and Watson (2008) emphasizing that learning is a movement from novice towards mature participation in the practices of communities. Extending Vygotsky's work, Rogoff (2008) has explored issues of apprenticeship, emphasizing the importance of *guided participation* during activities that are culturally valued. For children in less developed areas of the world, such experiences as engagement in farming, household or craft activities are likely to be common (e.g. Rogoff, 2003; Göncü, 1999); but in the west such undertakings are less often experienced and parents' work outside the home is likely to be unseen and largely unknown.

As early as 1995, Gifford suggested that apprenticeship in learning mathematics will support children's understandings of notations. Rogoff (2003) proposes that where young children either observe or engage with their parents work at home or elsewhere, that their *intent participation* in the adult's activity involves a combination of observing, taking the initiative and sensitive support which she compares to 'mastery learning' in apprenticeship (2003, p. 323). In the current study, analysis showed it was the children's cultural knowledge that most influenced the *children's interests*, their *intent participation* (observing and 'listening in') encouraging them to explore further (Rogoff et al., 2003).

Following publication of research by Moll, Amanti, Neff and Gonzalez (1992) into children's 'funds of knowledge' for their learning, Riojas-Cortez investigated role-play showing 'how culture is intertwined' (2001, p. 39), and recent research showed that children's 'funds of knowledge' explored in their impromptu play episodes includes their cultural knowledge of mathematics (Worthington & van Oers, 2016). Hedges, Cullen and Jordan (2011) list children's participation in cultural experiences at home, including household tasks and shopping; in a parent's work; in their parents' interests and leisure activities and in their siblings and grandparents' activities. In the present study, the first two contexts appeared to provide the richest cultural knowledge, with a parent's work appearing to be the most powerful: this was particularly marked for Isaac.

13.4.1 Ayaan's 'Funds of Knowledge'

Ayaan had only started attending the nursery school at the beginning of the academic year in which the data were gathered. Both her mother and teacher commented that she was very shy, and when she first came to nursery she only spoke Somali. Ayaan's father is a taxi driver and when he returns home each day Ayaan and her brothers and sisters surround him, clamouring to count his takings. They ask how many hours he's been at work, telling him they'll give him money (peas) to stay at home. Ayaan often goes shopping with her mother and siblings and loves helping prepare food at home. Ayaan was keen to learn Arabic like her brother, and her aunt agreed to teach her, using a large blackboard in the corner of their sitting room. Ayaan also loves watching television and helps care for her younger siblings. Most of Ayaan's pretend-play episodes focused on playing 'ice cream shops'.

13.4.2 Isaac's 'Funds of Knowledge'

Isaac's father was a builder, and Isaac had been involved in a great deal of his work and the conversion of their home, together with all aspects of joinery, measuring, wiring, plumbing and construction. Isaac's dad had a large toolbox in the kitchen, and Isaac had his own workbench and tools there. Additionally, Isaac's interests (and his involvement in his father's work) included many technologies such as security (e.g. surveillance cameras and padlocks) and mains services such as gas, power and water, and for more than two years Isaac's favourite bedtime reading had been a builders' trade catalogue. Isaac's father now owns and runs a microbrewery, work that includes Isaac helping with deliveries, payments, invoices and cash.

Isaac often goes camping with his father. His father is also very interested in maps and had hung some large old maps of Britain and New Zealand on the walls of their home. Together they often pored over the details of contemporary maps, relating them to the city in which they lived and journeys they'd make. At home, Isaac had drawn his own map that was stuck on the wall of the stairs. During a home visit (made by the author), Isaac pointed out the location of their home and his nursery to the author (during a home visit), on the swirling lines he'd drawn. Sometimes Isaac stays at his grandfather's house in the country, where his grandfather shares his knowledge of the steam trains that run nearby. During the year, Isaac initiated and engaged in pretend-play more than any other child in the study; his play was highly complex and often sustained. All Isaac's many play narratives related to his personal 'funds of knowledge'.

13.4.3 Shereen's 'Funds of Knowledge'

In Shereen's family buying, preparing, ordering and eating food together (sometimes in a café) are highly valued as significant social activities. On one occasion, Shereen's grandmother had travelled to England from the Philippines to visit her daughter's family, and making a home visit at this time, I was offered a drink. A few minutes later, Shereen's grandmother came out of the kitchen and gave me a plate of noodles. It was easy to appreciate that sharing food was a part of the wider family's values too. All of Shereen's pretend-play narratives centred on food, either shopping for food (writing shopping lists, receipts, open and closed signs) or playing cafes in which she took orders for food, informed customers of items available and their cost and admonished those who failed to eat everything.

13.5 Pretend-Play as a Context for Meaningful Exploration of Children's Interests

Play has been the subject of numerous research studies although there is no agreement on its role in the curriculum, and according to Smith its functions continue to be a focus of debate, and the extent to which it is valued culturally and its frequency varies (Smith, 2010; Corsaro, 2005). Vygotsky equated 'play' with *pretend-play*, 'the leading factor in development' in the years before school (1978, p. 101).[2] In Vygotsky's view, pretend-play can provide rich contexts for children to explore their cultural knowledge, helping create a 'bridge' between spontaneous (everyday) concepts and scientific concepts later in school. Vygotsky also highlighted children's ability to ascribe alternative meanings to objects in pretend-play, underpinning our ability to use graphical signs to signify meanings (1978).

Pretend-play is widely acknowledged as an important aspect of early childhood education (e.g. OECD, 2006) and is included in influential curriculum frameworks internationally such as *Te Whariki* in New Zealand (Ministry of Education, 1996). Pretend-play fulfils many roles and can enable children to imitate and extend home cultural experiences in which they have been apprenticed (Rogoff, 2008). For example, Garrick (2012) refers to ethnographic research in Italy and the USA 'noting the processes of young children's exploration of shared concerns and interests from their social worlds, as they engage in the reconstruction and interpretation of past events' (Garrick, 2012, p. 3). Findings from research by Gmitrova and Gmitrov showed that 'child-directed play… evoke[s] a balanced relationship between the cognitive and affective sphere' (2003, p. 245), showing that:

> … the lowered rate and level of cognitive domain behaviors from children during the teacher's direction of play is related to the poorer use of natural drive of the pretend-play in the education process. During adult-planned play, 'children's persistence gradually decreased, thereby shortening the duration [of play] compared with the free play condition, when children take greater pleasure in play and learning (2003, pp. 245–246).

Broadhead and Burt (2012) and Drummond and Jenkinson (2009) highlight the value of 'more flexible and loosely defined spaces, places and things' for pretend-play (Garrick, 2012, p. 1). However, a number of studies (e.g. Gifford, 2005; Brannon & van de Walle, 2001; Ewers-Rogers & Cowan, 1996; Munn & Schaffer, 1993) found little evidence of mathematics in pretend-play (Worthington & van Oers, 2016), and a systematic study of children in ten early childhood settings in England confirmed these findings (Moyles & Worthington, 2011). In many of these settings designated adult-planned role-play 'areas' have replaced genuine child-initiated play, Worthington and van Oers highlighting 'practice common in most of the world where adults choose, plan and resource themed role-play areas, revealing *adults' perceptions* of children's interests' (2016, p. 52).

---

[2]That is until 7 years of age when 'school' learning of *scientific concepts* take precedent (Vygotsky).

In this nursery school, the children's play is spontaneously generated, collaborative, imaginative and creative, allowing complex explorations of their worlds. Much of the children's play is only paused by a need to stop for lunch or to go home: in such instances, children frequently continue their play narratives over several days, sometimes returning to the same theme during more than one term (Worthington, 2010). Emma, one of the teachers explained, '*Longer play periods where children choose to extend their play ideas, requires the adults to see play not only in short periods or through a day but something children revisit again and again*'. The three children's play narratives included in this chapter are both short and sustained, chosen to exemplify the oral language and their *mathematical graphics*.

13.6 The Children's Mathematical Graphics

Spontaneous pretend-play allows children to explore their cultural knowledge without forfeiting ownership. It has the potential to support children's meaning making and communication through graphicacy (i.e. drawing, maps, writing and *children's mathematical graphics*) (Worthington & van Oers, 2015) and plays an important role in their subsequent understanding and use of increasingly formal school mathematical notations (Worthington, 2012; Carruthers & Worthington, 2006). From this perspective understanding, the cultural knowledge of a discipline involves 'psychological activity' with sign systems that are subsequently 'reconstituted and developed to form a new psychological entity' (Vygotsky, 1978, p. 57). The implication is that mathematical understandings and the signs children use in early childhood change when they meet the formal abstract symbolic language of mathematics in school.

Researching children's graphic development Machón (2013) found that between the ages of three and four years children begin to experiment and use distinct graphic symbols. This finding is mirrored in a recent study of children's social literacy (Worthington & van Oers, 2015). Throughout this research, the term 'marks' is used to refer to scribble-marks, while the term 'symbol' refers to graphical symbols such as arrows, circles and hearts. Some terms (such as crosses) are often used interchangeably to signify alternative meanings that are dependent on their context (Worthington, 2009). The term 'sign' is used in this research to refer writing-like marks such as wavy lines, alphabetic letters, numerals and other mathematical signs (e.g. £ and +). In their play narratives, the children employed a range of mathematical genres including shopping lists, orders in cafes, registers, cheques, receipts, environmental signage, goals scored in various football matches and booking for a campsite.

13.6.1 Documented Observations of Children's Pretend-Play

Investigating the cultural foundations of mathematics at home highlighted the relationship between children's cultural home experiences and their social pretend-play narratives (Worthington & van Oers, 2016). In over 40% of their pretend-play episodes, the children included some reference (oral and graphical) to mathematics, with variation among individuals. References to *number, quantities and counting* appeared most frequently, followed by *time* and *money*; these aspects appeared to be the areas of mathematics in greatest use in their homes. The children's play also revealed their cultural knowledge of length and distance, direction, speed, weight, temperature, shape, space and capacity and data handling—some aspects extending beyond the early years curriculum in England (DfE, 2014). None of the children had received any formal instruction in mathematics in their nursery school. Worthington and van Oers observe that:

> …the children's cultural knowledge of home and community permeated their pretence in a functional way, melding their lived experiences and cultural knowledge with imagined possibilities and underscoring that meaning is the foundation of semiotic behaviours. The children chose to be involved in emulations of everyday practices that required communication (2015, p. 17).

The following examples highlight the children's communications through their *mathematical graphics* and represent the range of graphical marks, signs and symbols they used at the time; their play narratives allowing further refinement and progression of both their graphics and mathematical understandings.

Participating in the children's play when invited, the children's teacher supported their play and mathematical thinking through 'observing and taking an interest, providing resources to support children's interests, creating mathematically and graphically rich environments, modelling graphics for authentic purposes throughout each day and mediating learning through collaborative dialogue about the children's thinking and representations' (Worthington & van Oers, 2016, p. 63).[3]

13.6.2 Play Narrative 2

Isaac wrote '*A cheque for £500.00, for all the jobs I've done at my house*' (Fig. 13.2).

Writing zigzag lines on a play cheque book followed by a letter-like 'a', Isaac had used *early explorations with marks, attaching mathematical meanings* to them, and *early written numerals*. Young children often use writing-like marks (zigzag or

---

[3]Modelling the use of graphical symbols, signs and texts for authentic purposes is an essential feature of children's mathematical graphics.

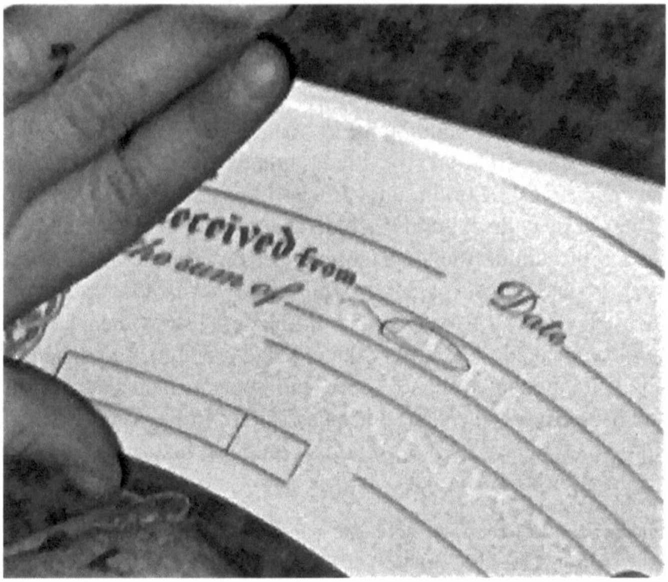

Fig. 13.2 Isaac's cheque

wavy lines) to signify their meanings. For example, Machón (2013) found that children appear to notice and imitate adults' linear arrangements of writing as beginning around the age of three years. This example (and Ayaan's in play narrative 3) shows that young children sometimes use letters and numerals interchangeably.

13.6.3 Play Narrative 3

Playing cafés outside with friends, Shereen took orders for food. She used vertical lines as shorthand for numbers of items on her list and counted each line with one-to-one correspondence up to 20 (Fig. 13.3).

Shereen *represented quantities that she counted*. Her teacher had not directly taught the children to use tallies, though may have modelled their use in meaningful contexts.

13.6.4 Play Narrative 4

For two weeks Ayaan had been playing in the gazebo, offering pretend ice cream through the window to children. Today when a child replied '*Yes*', Ayaan answered

Fig. 13.3 Shereen's order

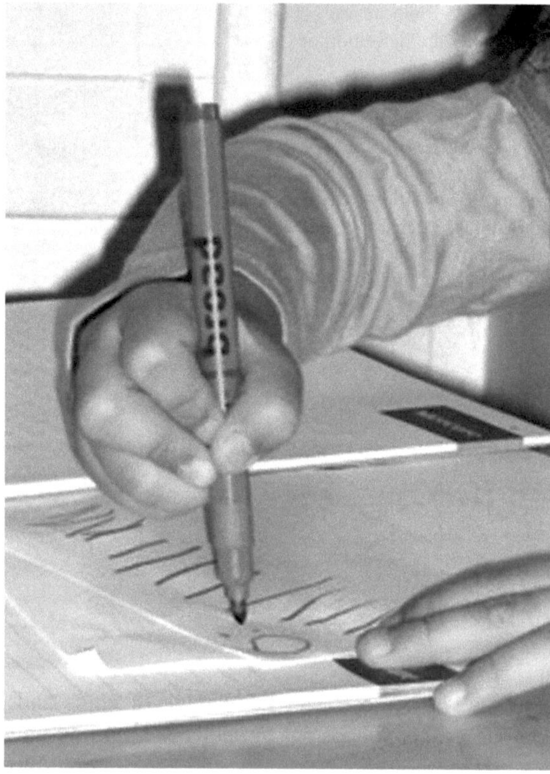

'*No left*', adding '*I make more*'. Collecting stones and pretending to make ice cream, Ayaan asked Tariq if he wanted any, and writing a small letter 'A' she passed him an imaginary one, then pressed buttons on the till saying, '*It's 50 min*' (Fig. 13.4). Shortly afterwards, Ayaan drew dashes in a notebook without comment. The next time Ayaan played ice cream shops she asked '*50 min please*'. When a child offered '*£1.00*' Ayaan replied '*That's £50.00 please*'.

In this episode, Ayaan drew on her growing knowledge of English vocabulary, trying out mathematical terms relating to time and money. Ayaan also communicated through her graphics, using *early explorations with marks, attaching mathematical meanings* to them, and *early written numerals*. Carruthers and Worthington use the term 'scribble-marks' not in any derogatory sense, but when referring to those marks that adults would find difficult to understand without the child's explanation. The children often seemed to use scribble-marks within their play as 'placeholders' to denote specific meanings, suggesting that such rapidly made marks allow the course of play to proceed uninterrupted' (Worthington & van Oers, 2015, p. 19). Ayaan's use of the initial letter of her name is also very powerful in terms of her personal identity and was something she had written in her play on several occasions. During the year, Ayaan's confidence and use of English grew, her teacher commenting that her pretend-play appeared to contribute considerably to both.

Fig. 13.4 '50 min please (Regrettably the definition of this photograph is insufficiently sharp to show the 'A' Ayaan wrote)', Moyles, *The Excellence of Play*, 2015. Reproduced with the kind permission of Open International Publishing Ltd.

13.6.5 Play Narrative 5

Drawing on her personal knowledge of cafés and imitating a waitress, Shereen approached her friends for 'orders', making lines and other writing-like marks on a notepad as she did so. After a while she returned to ask her teacher Emma, '*what you want: rice, chocolate, cake, chicken?*' Emma said she didn't want chicken and Shereen wrote a mark for 'chicken' and drew a cross by it, clarifying, '*It says cross ['x']—no chicken*'. Later Emma said she would have chicken, but pointing to the 'x' she had written, Shereen said '*Look! No chicken! You want mushroom?*' Then pointing to her drawing of a mushroom explained '*Look. A tick, that mean we got some*' (Fig. 13.5).

Shereen's understanding of the meanings of crosses and ticks was unambiguous. She used a combination of *early explorations with marks, attaching mathematical meanings,* and *explorations with symbols*. This example is unusual in that Shereen combined several graphical forms and included drawings (a fish and a mushroom). Previous research by Carruthers and Worthington (2006) showed that children often *code-switch* within one graphical text, beginning with one mark, sign or symbol and changing to another, enabling them to best signify their meanings.

Fig. 13.5 'You want mushroom? Look! A tick—that mean we got some', Carruthers and Worthington, *Understanding Children's Mathematical Graphics*, 2011. Reproduced with the kind permission of Open International Publishing Ltd. All rights reserved

13.7 Implications and Likely Trajectories

This research reveals the beginnings of abstraction in early childhood mathematics, the children's interests underpinning their subject knowledge of mathematics. The findings have pedagogical implications for all those working with young children. Emphasizing the cultural and transformational nature of children's textual trajectories, this chapter argues that their existing competencies and informal representations be prioritized.

Policies for early childhood education in England are seriously out of step with what young children need to support effective mathematical understandings. The current situation is exacerbated by an increasingly formal start in school at four years of age, a start that now severely marginalizes play, privileging calculations even in nursery and inhibiting children's mathematical thinking. Educators in England have urged that curricula and pedagogy be based on evidence from research, and the school entry age be changed to seven years in keeping with most other European countries.

Practical pedagogical developments relating to children's abstraction grow through the expediency of teachers' desire to make a genuine difference in mathematics in their nursery schools and schools and having the freedom to do so (see for example, Carruthers, 2015). Teachers can really make a difference, especially when they collaborate with colleagues to share ideas and develop their understandings and practice. As more researchers and teachers in England and elsewhere explore the potential of children's informal mathematics, the more can be understood. Children's mathematics has the potential to become the focus of pedagogy throughout this phase, provided early childhood education in England can regain its child-centered focus appropriate to young children's learning and development. Interest in more open ways of young children's mathematics includes many European countries (e.g. the Netherlands, Germany, Portugal and Spain), Australia, New Zealand, the Russian Federation and Ukraine, suggesting that there is greater potential for its development.

The findings of this study and the body of research on which it builds, contribute to the increasing scholarship and theoretical trajectories concerning children's abstract thinking and symbolic representation in early childhood mathematics (e.g. Gravemeijer et al., 2002). Further research is needed into the relationship between children's informal representations and 'formal' mathematical codes and the impact of young children's informal mathematics on later outcomes including calculations.

13.8 Summary and Conclusion

The findings presented in this chapter show that children's early understandings of the abstract symbolic language of mathematics are social, contextual and communicative, their cultural knowledge transformed through their interests explored within their social pretend-play. When Ayaan, Shereen and Isaac drew on their cultural knowledge of maps, builders, paying bills, writing orders in a café and selling ice creams, they were actively involved in culturally meaningful mathematics and engaged in abstract thinking. The findings make a strong case for prioritizing learners' own mathematics to ensure that young children's mathematical semiosis meshes with the increasingly formal abstract mathematical notations in school.

Of some considerable concern is the highly politicized educational agenda in England that increasingly results in the '*pedagogization* of play', 'planned' or 'structured' to meet specific curriculum targets, (Rogers, 2010, p. 163, emphasis in the original). This impacts on young children's experiences so that play is often marginalized and misunderstood (Moyles & Worthington, 2011). Wood emphasizes;

> … adults' plans and purposes are privileged [distorting] the more complex meanings and purposes of play, including the ways in which children exercise power, agency and control, how they make and communicate meanings through symbolic activity, and the significant role of representations in play (2010, p. 18).

Drawing on their home cultural knowledge, the children in this study imitated and extended their interests, their use of spontaneous mathematical concepts making sense within their developing narratives (Worthington & van Oers, 2016). Children's own mathematical representations are *meaningful*, and while their thinking and graphical communications may differ considerably from standard 'school' mathematics, their cultural knowledge and their marks, symbols and mathematical texts deserve our respect for the very considerable understandings they reveal.

Rather than referring to mathematics only as a subject or curriculum, supporting *children's interests* points to the centrality of their mathematical knowledge and thinking (e.g. Carruthers & Worthington, 2005, 2011; DCSF, 2009). This chapter challenges current views of a single 'skills-based' mathematics and of 'planned play' showing how spontaneous and 'sustained social pretend-play can create a rich social-ecocultural niche in early childhood' (Worthington, 2015a). This perspective unsettles current political-driven agendas, narrow curriculum targets and the often context-free mathematical notations familiar in England and in many countries, suggesting instead a positive perspective from which children's mathematical literacy and agency can be understood, and a paradigm shift for early childhood mathematics.

Acknowledgements Sincere thanks go to the children, parents, staff and headteacher of Redcliffe maintained nursery school, Bristol, for sharing the children's inspirational play and mathematical thinking.

References

Aubrey, C. (1993). An investigation of the mathematical knowledge and competencies which young children bring into school. *British Educational Research Journal, 19*(1), 27–41.

Aubrey, C. (1997). *Mathematics teaching in the early years*. London: Falmer Press.

Bertram, T., & Pascal, C. (2002). What counts in early learning? In O. N. Saracho & B. Spodek (Eds.), *Contemporary perspectives on early childhood curriculum* (pp. 241–259). Greenwich, CT: Information Age Publishing.

Brannon, E. M., & van de Walle, G. (2001). The development of ordinal numerical knowledge in young children. *Cognitive Psychology, 43*(1), 53–81.

Broadhead, P., & Burt, A. (2012). *Understanding young children's learning through play*. Abingdon, Oxon: Routledge.

Brooker, L. (2010). Learning to play, or playing to learn? In L. Brooker. & S. Edwards. (Eds.), *Engaging play* (pp. 39–52). Maidenhead: Open University Press.

Carr, M. (2001). *Assessment in early childhood settings: learning stories*. London: Sage Publications.

Carruthers, E. (1997). *Number: a developmental theory: a case study of a child from 20 to 44 months*. Unpublished Masters (M. Ed.) Dissertation. University of Plymouth.

Carruthers, E. (2015). Listening to children's mathematics in school. In B. Perry., A. Gervasoni., & A. MacDonald. (Eds.), *Mathematics and transition to school—International perspectives* (pp. 313–330). Sydney, Australia: Springer.

Carruthers, E., & Worthington, M. (2005). Making sense of mathematical graphics: the development of understanding abstract symbolism. *European Early Childhood Education Research Journal, 13*(1), 57–79.

Carruthers, E., & Worthington, M. (2006). *Children's mathematics: Making marks, making meaning*. London: Sage Publications.

Carruthers, E., & Worthington, M. (2008). Children's mathematical graphics: calculating for meaning. In I. Thompson, (Ed.), (2008) *Teaching and Learning Early Number*. Maidenhead: Open University Press, (2nd ed.).

Carruthers, E., & Worthington, M. (2011). *Understanding children's mathematical graphics: Beginnings in play*. Maidenhead: Open University Press.

Carruthers, E., & Worthington, M. (2013). *Taxonomy charting children's mathematical graphics.* http://www.childrens-mathematics.org.uk/taxonomy.pdf.

Christie, J., & Roskos, K. (2009). Play's potential in early literacy development. *Encyclopedia on Early Childhood Development.* http://www.childencyclopedia.com/sites/default/files/dossiers-complets/en/play.pdf#page=31.

Clayden, E., Desforges, C., Mills, C., & Rawson, W. (1994). Authentic activity and learning. *Journal of Education Studies, 42*(2), 163–173.

Corsaro, W. (2005). *The sociology of childhood.* London: Sage.

Curtis, A. (1998). *A curriculum for the preschool child* (2nd ed.). London: Routledge.

David, M. M., & Watson, A. (2008). Participating In what? Using situated cognition theory to illuminate differences in classroom practices. In A. Watson, A., & P. Winbourne (Eds.), *New directions for situated cognition in mathematics* (pp. 31–56). London: Springer.

DCSF. (2009). *Children thinking mathematically: PSRN essential knowledge for early years practitioners.* London: DCSF. http://childrensmathematics.net/childrenthinkingmathematically_psrn.pdf.

DfE. (2013). *Early years foundation stage profile: Handbook.* London: Standards and Testing Agency, https://www.gov.uk/government/uploads/system/uploads/attachment_data/file/301256/2014_EYFS_handbook.pdf.

DfE. (2014). *Statutory framework for the early years foundation stage.* http://dera.ioe.ac.uk/19831/1/DFE-00337-2014.pdf. Accessed 5th July 2016.

Drummond, M. J., & Jenkinson, S. (2009). *Meeting the child. Approaches to observation and assessment in Steiner kindergartens*. Plymouth: The University of Plymouth.

Empson, S. B., & Jacobs, V. R. (2008). Learning to listen to children's mathematics. In D. Tirosh & T. Wood (Eds.), *The international handbook of mathematics teacher education*: Vol. 2. *Tools and processes in mathematics teacher education* (pp. 257–281). Rotterdam, The Netherlands: Sense Publishers.

Ewers-Rogers, J., & Cowan, R. (1996). Children as apprentices to number. *Early Childhood Development and Care, 125*(1), 15–25.

Fleer, M. (2010). *Early learning and development: cultural-historical concepts in play.* Cambridge: Cambridge University Press.

Garrick, R. (2012). Pretend play: the affordances of flexible spaces, places and things for an interest based curriculum. Draft paper presented at the *British Educational Research Association Annual Conference*, University of Manchester, 4–6 September 2012.

Gifford, S. (1995). Number in early childhood. *Early Child Development and Care, 109*(1), 95–119.

Gifford, S. (2005). *Teaching mathematics 3-5*. Maidenhead: Open University Press.

Gmitrova, V., & Gmitrov, J. (2003). The impact of teacher-directed and child-directed pretend play on cognitive competence in kindergarten children. *Early Childhood Education Journal, 30*(4), 241–246 (Summer).

Göncü, A. (Ed.). (1999) *Children's engagement with the world.* Cambridge: Cambridge University Press.

Gravemeijer, K., Lehrer, R., & van Oers, H. (2002). *Symbolizing, modeling and tool use in mathematics Education*. Dordrecht, The Netherlands: Kluwer.

Hedges, H., & Cullen, J. (2005). Subject knowledge in early childhood curriculum and pedagogy: Beliefs and practices. *Contemporary Issues in Early Childhood, 6*(1), 66–78.

Hedges, H., Cullen, J., & Jordan, B. (2011). Early years curriculum: funds of knowledge as a conceptual framework for children's interests. *Journal of Curriculum Studies, 43*(2), 185–205.

Hughes, M. (1986). *Children and number: Difficulties in learning mathematics*. Oxford: Basil Blackwell.

Lakoff, G., & Nunez, R. (2001). *Where mathematics come from: How the embodied mind brings mathematics into being*. New York: Basic Books.

Lave, J., & Wenger, E. (1991). *Situated learning: Legitimate peripheral participation*. Cambridge: Cambridge University Press.
Machón, A. (2013). *Children's drawings: The genesis and nature of graphic representation*. Madrid: Fibulas Publishers.
Marcon, R. (2002). Moving up the grades: Relationships between preschool model and later school success. *Early Childhood Research & Practice, 4*(1). http://www.ecrp.uiuc.edu/v4n1/marcon.html.
Ministry of Education. (1996). *Te Whäriki. He whäriki mätauranga mō ngä mokopuna o Aotearoa:* Early Childhood Curriculum. Wellington: Learning Media. http://www.education.govt.nz/early-childhood/teaching-and-learning/ece-curriculum/te-whariki/.
Moll, C., Amanti, C., Neff, D., & Gonzalez, N. (1992). Funds of knowledge for teaching: using a qualitative approach to connect homes and classrooms. *Theory Into Practice, 31*(2), 132–141.
Moyles, J., Adams, S., & Musgrove, A. (2002). *SPEEL study of pedagogical effectiveness in early learning*. http://www.dfes.gov.uk/research/data/uploadfiles/RR363.pdf.
Moyles, J., & Worthington, M. (2011). *The early years foundation stage through the daily experiences of children' TACTYC Occasional Paper No. 1*. http://tactyc.org.uk/occasional-paper/occasional-paper1.pdf.
Munn, P. (1994). The early development of literacy and numeracy skills. *European Early Childhood Education Research Journal, 2*(1), 5–18.
Munn, P., & Kleinberg, S. (2003). Describing good practice in the early years—A response to the 'third way'. Education 3-13. *International Journal of Primary, Elementary and Early Years Education, 31*(2), 50–53.
Munn, P., & Schaffer, R. (1993). Literacy and numeracy events in social interactive contexts. *International Journal of Early Years Education, 1*(3), 61–80.
National Agency for Education. (2011). *Curriculum for the Preschool Lpfö 98, Revised 2010*. Stockholm: Frizes. http://skolverket.se/publikationer?id=2704.
Nutbrown, C. (1999). *Threads of thinking* (2nd ed.). London: Paul Chapman.
OECD. (2006). *Starting strong II: Early child education and care*. Paris, France: OECD Publishing. http://www.unicef.org/lac/spbarbados/Implementation/ECD/StartingStrongII_OECD_2006.pdf.
Poland, M. (2007). *The treasures of schematising. Doctoral dissertation*. Amsterdam: VU University.
Poland, M., van Oers, B., & Terwel, J. (2009). Schematising activities in early childhood Education. *Educational Research and Evaluation, 15*(3), 305–321.
Pramling, I., & Johansson, E. (2006). Play and learning—inseparable dimensions in preschool practice. *Early Child Development and Care, 176*(1), 47–65.
Riojas-Cortez, M. (2001). Preschoolers' funds of knowledge displayed through sociodramatic play episodes in a bilingual classroom. *Early Childhood Education Journal, 29*(1), 35–40.
Rogers, S. (2010). Powerful pedagogies and playful resistance. In L. Brooker & S. Edwards (Eds.), *Engaging play* (pp. 152–165). Maidenhead: Open University Press.
Rogoff, B. (2003). *The cultural nature of human development*. Oxford: Oxford University Press.
Rogoff, B. (2008). Observing sociocultural activity on three planes. In P. Murphy, K. Hall & J. Soler. (Eds.), *Pedagogy and practice: Culture and identities* (pp. 58–74). London: Sage Publications.
Rogoff, B., Paradise, R., Mejia Arauz, R., Correa-Chávez, M., & Angelelillo, C. (2003). Firsthand learning through intent participation. *Annual Review of Psychology, 54,* 175–203.
Smith, P. (2010). *Children and play*. Chichester: Wiley-Blackwell.
Street, B., Baker, D., & Tomlin, A. (2008). *Navigating numeracies: Home/school numeracy practices*. London: Springer.
Tizard, B., & Hughes, M. (1984). *Young children learning: Talking and thinking at home and at school*. London: Fontana.
van den Heuvel-Panhuizen, M., & Drijvers, P. (2014). Realistic mathematics education. In S. Lerman (Ed.), *Encyclopedia of mathematics education* (pp. 521–525). Dordrecht, Heidelberg: Springer.

van Oers, B. (2012). How to promote young children's mathematical thinking? *Mediterranean Journal for Research in Mathematics Education, 11*(1–2), 1–15.

van Oers, B. (2013). Communicating about number: Fostering young children's mathematical orientation in the world. In L. English & J. Mulligan (Eds.), *Reconceptualising early mathematics learning* (pp. 183–203). New York: Springer.

Vygotsky, L. S. (1978). *Mind in society: The development of higher psychological processes.* Cambridge, MA: Harvard University Press.

Vygotsky, L. S. (1998). *Child psychology.* In L. S. Vygotsky. *The collected works of L. S. Vygotsky. Vol. 5.* New York: Kluwer Academic and Plenum Publishers.

Wood, E. (2010). Reconceptualizing the play-pedagogy relationship. In L. Brooker & S. Edwards (Eds.), *Engaging play* (pp. 11–24). Maidenhead: Open University Press.

Worthington, M. (2009). Fish in the water of culture: Signs and symbols in young children's drawing. *Psychology of Education Review, 33*(1), 37–46.

Worthington, M. (2010). Play is a complex landscape: Imagination and symbolic meanings. In P. Broadhead, L. Wood. & J. Howard. (Eds.), *Play and learning in educational settings* (pp. 127–144). London: Sage Publications.

Worthington, M. (2012). Children becoming expert symbol users. In M. McAteer (Ed.), *Improving primary mathematics teaching and learning* (pp. 39–57). Maidenhead: Open University Press.

Worthington, M. (2015a). Mathematics and the ecology of pretend play. In J. Moyles. (Ed.), *The excellence of play* (pp. 237–249). 4th Ed. Maidenhead: Open University Press.

Worthington, M. (2015b). Young children's informal mathematical signs and symbols: cultural learning and 'intent participation'. *Presentation at the European Early Childhood Education Research Association* (EECERA) conference. Barcelona, 07–10 September.

Worthington, M., & Carruthers, E. (2003). *Children's mathematics, making marks, making meaning*. London: Paul Chapman Publishing.

Worthington, M., & van Oers, B. (2015). Children's social literacies: Meaning making and the emergence of graphical signs and texts in pretence. *Journal of Early Childhood Literacy. Published online December 2015.* http://ecl.sagepub.com/content/early/2015/12/09/146879841 5618534.full.pdf+html.

Worthington, M., & van Oers, B. (2016). Pretend play and the cultural foundations of mathematics. *European Early Childhood Education Journal, 24*(1), 51–66.

Worthington, M., & van Oers, B. (Forthcoming). *The emergence of abstraction in the nursery.* Educational Studies in Mathematics.

Author Biography

Maulfry Worthington has lectured in early childhood and mathematics education in the university sector, worked as a National Numeracy Consultant and has extensive experience in teaching young children. She currently teaches on Masters' level early years mathematics courses. Maulfry's research interests include children's mathematics, graphicacy and pretend-play. She has conducted extensive research individually and jointly with Elizabeth Carruthers into young children's mathematical notations and their book, Children's Mathematics: Making Marks, Making Meanings is acknowledged as a seminal text. Maulfry has published widely and is currently completing her doctoral research (VU University, Amsterdam), into the genesis of mathematical semiosis in early childhood. Maulfry and Elizabeth are founders of the international Children's Mathematics Network.

Chapter 14
Making Connections Using Multiplication and Division Contexts

Jennifer Young-Loveridge and Brenda Bicknell

Abstract This chapter focuses on the connections that can be made by using multiplication and division problem-solving contexts in early childhood education and school settings. Prior to starting school, young children experience many opportunities to make groups using familiar objects, beginning with groups of two and then moving to larger groups such as five and ten. Typically, children begin by using units of one, as shown in counting one-by-one. However, children should experience "groups of" objects larger than one (composite units) early on in their schooling. Another key idea for children to understand is the concept of additive composition, the way that numbers are composed of other numbers (part–whole relationships). The connections are explored between mathematics learning in informal and formal settings; ordinality and cardinality; composing and decomposing quantities; operations and processes; and word problems and representations. To illustrate these connections, we draw on a two-year study undertaken with 84 culturally and linguistically diverse five- to eight-year-olds. During the study, children participated in a series of lessons where they solved multiplication and division problems involving naturally occurring groups of twos, fives, and tens using a variety of materials and multiple representations. Results for the 35 five-year-olds showed improvement in number knowledge, addition and subtraction, early place-value understanding, as well as multiplication and division.

Keywords Counting · Multiplication and division · Representations
Materials · Place value · Early years of school

J. Young-Loveridge (✉) · B. Bicknell
The University of Waikato, Hamilton, New Zealand
e-mail: educ2233@waikato.ac.nz

B. Bicknell
e-mail: bicknell@waikato.ac.nz

V. Kinnear et al. (eds.), *Forging Connections in Early Mathematics Teaching and Learning*, Early Mathematics Learning and Development,
https://doi.org/10.1007/978-981-10-7153-9_14

14.1 Background Literature

The concept of a unit is a fundamental idea underpinning all of mathematics learning. Many researchers have written about the idea of the unit in mathematics (e.g., Behr, Harel, Post, & Lesh, 1994; Confrey & Harel, 1994; Lamon, 1994, 1996; Sophian, 2007; von Glasersfeld, 1995). Behr and colleagues (1994) claim that there is a hidden assumption pertaining to primary school mathematics, which is that "all quantities are represented in terms of units of one" (p. 123). Learning to count (by ones) is an example of the use of singleton units. Gelman and Gallistel's (1978) first counting principle focuses on the one-to-one relationship between the counting words and the objects being counted. Learning the sequence of number words in the correct order and being able to use this sequence in conjunction with objects to determine how many items are in a collection is important (Geleman & Gallistel, 1978). However, too much counting by ones is not necessarily desirable, particularly if it continues to be the strategy of choice for problem solving (Young-Loveridge, 2001).

Two other concepts thought to be important for children's mathematical understanding are *cardinality* (how many items altogether) and *ordinality* (what number comes before or after a particular number) (Sarama & Clements, 2009). Cardinality is sometimes described by "number as *hierarchical inclusion*" (i.e., each number includes all the numbers that come before it in the sequence of numbers). Ordinality has been described by "number as *seriation*" (i.e., each number comes after and is one more than the previous number in the sequence of numbers) (Sarama & Clements, 2009). There is a parallel in the distinction between *collections-based* and *counting-based* conceptions of number (Yackel, 2001). Experiences with counting help young children build knowledge of number sequence and ordinality. A collections-based approach focuses on the composition of numbers in terms of groups (set–subset or part–whole relationships), drawing on understanding of hierarchical inclusion and cardinality (Sarama & Clements, 2009). Yackel's (2001, p. 24) view is that "having access to both conceptions provides much more flexibility for students … Therefore, [it is important] to foster the development of both conceptions."

Although it has been argued that it is important for students to develop both counting-based and collections-based approaches to working with numbers (Yackel, 2001), Yang and Cobb (1995, p. 10) have highlighted "an inherent contradiction" in the way that Western children are initially encouraged to count by ones and thus construct unitary *counting-based* number concepts, which then need to be reorganized into *collections-based* concepts involving units of ten and one when place-value instruction begins. Yang and Cobb contrast the Western counting-based view with the collections-based approach of Chinese mothers and teachers, who emphasize groups (units) of ten. The difference in emphasis on counting versus grouping by tens helps to explain Yang and Cobb's (1995) finding of more advanced mathematical understanding by Chinese children compared to that of American children.

One of the challenges for children in learning to use part–whole thinking is in recalling number facts. Research shows that low achievers in mathematics have consistent difficulties in recalling number facts and using them to solve problems (Baroody, 2011). One of the reasons suggested for these difficulties is that learners often continue to rely on counting strategies, requiring a lot of energy and attention, meaning that number facts have less chance of becoming known facts (Gray, 1991). Several researchers have highlighted the importance of learners developing automaticity in mathematics (Gray, 1991; Hattie & Yates, 2014; Hopkins & Lawson, 2002). Some research has shown that young children can be taught how to derive new facts from those they know, without necessarily needing to have progressed from a "counting all" strategy to a "counting on" strategy (e.g., Fischer, 1990; Henry & Brown, 2008; Steinberg, 1985). It is suggested that learning to recall number facts and using them to solve new problems should be made more explicit (Baroody, 2011). This mastery of number facts grows out of a systematic approach emphasizing patterns and relations, which helps learners to recognize the interconnectedness among facts and operations (e.g., $3 + 3$ is the same as 2×3, which is double three, and double-double three is $6 + 6$ which is 4×3 which is 12) (Baroody, Bajwa, & Eiland, 2009).

14.2 Curriculum Reform

Children begin to think mathematically and use "powerful mathematical ideas" from an early age (Clements & Sarama, 2007, 2009; Perry & Dockett, 2008), and this is reflected in current curriculum for children in the early years. In early childhood education, there is a greater focus on "interactions, experiences, routines and events, planned and unplanned, that occur in an environment designed to foster children's learning and development" (Commonwealth of Australia, 2009, p. 9). These principles reflect current early childhood philosophy and pedagogical practices focused on the holistic nature of teaching and learning (see, e.g., Australia: Commonwealth of Australia, 2009; England: Department for Education, 2014; New Zealand: Ministry of Education, 1996; USA: National Association for the Education of Young Children [NAEYC], 2009, 2010). A variety of theories inform these documents, for example, sociocultural theories that emphasize mathematics as a cultural tool and constructivist, ecological, behavioral, and critical theory (Cobb, 2007). Mathematics has been approached through different aspects of development, including cognitive, linguistic, social-emotional, and physical, and is woven into the children's experiences in the early childhood setting (NAEYC, 2010).

Children enter formal schooling at differing ages, depending on the country. In some countries, children enter school before the age of five years (England) or on their fifth birthday (New Zealand), whereas in other countries it is more common to begin formal schooling at the age of about six (Australia, USA). Curricula designed for young children attending early childhood education centers in New Zealand tend to take a holistic approach to learning and development (Ministry of

Education, 1996). Curriculum documents intended for young children in formal schooling usually position mathematics as a separate curriculum area with specific mandated learning outcomes. The transition to school presents challenges in mathematics learning so that "this new stage in children's learning builds upon and makes connections with early childhood learning and experiences" (Ministry of Education, 2007, p. 41). Curriculum standards in mathematics (e.g., Ministry of Education, 2009) have placed greater emphasis on accountability and measurable outcomes, particularly in number.

Many mathematics curricula focus initially on counting and then on addition and subtraction, before introducing other domains such as multiplication and division, and proportional reasoning (e.g., Australia: ACARA, 2011; England: Department for Education, 2013; Ireland: NCCA, 1999; New Zealand: Ministry of Education, 2007; USA: CCSSI, 2010). However, research has shown that young children have considerable knowledge of multiplication and multiplicative relationships including division prior to formal instruction in this domain (e.g., Bakker, van den Heuvel-Panhuizen, & Robitzsch, 2014). Moreover, Bakker et al. found at this early stage there was no difference in difficulty between multiplication and division. Similarly, children are able to work with division as a sharing process (e.g., Davis & Pitkethly, 1990; Frydman & Bryant, 1988; Squire & Bryant, 2003).

In the New Zealand context, the achievement objectives for the first years of school make no explicit reference to multiplication and division, and this suggests that few teachers use multiplication and division contexts with young children. However, by about eight or nine years of age, students are expected to understand place value, knowing "how many ones, tens, and hundreds are in whole numbers to at least 1000" and to "know simple fractions in everyday use" (Level Two: Ministry of Education, 2007). This is somewhat paradoxical given the fact that place value is inherently multiplicative in its use of groups and powers of ten, as well as the fact that fractions are based on division processes. It is not until Level Three (Years 5–6, ages 9–11) that there is explicit reference to multiplication and division strategies, facts, and properties. This is similar to other education systems. For example, children in the USA are expected to understand place value in Grade 1 (age 6), despite the fact that the foundations for multiplication are not introduced until Grade 2 (age 7) (CCSSI, 2010). Multiplication only becomes a serious focus at Grade 3 (age 8). Similarly, in Ireland, place value is mentioned for first and second classes (ages 7–8), while multiplication and division are not mentioned until third and fourth classes (ages 9–10). In Australia and England, place value is part of the label "number and place value" for a domain that begins at Year 1 (England: Department for Education, 2013) or prior to Year 1 (Australia: ACARA, 2011) and continues through all years of the primary school system.

In recent decades, international mathematics education reform has called for a greater focus on building conceptual understanding within the context of mathematics problem solving, instead of the traditional approach to teaching mathematics by emphasizing memorization of procedures, facts, and rules (e.g., Buchholz, 2004; Skemp, 1978). The reform approach to mathematics teaching stresses the importance of thinking, reasoning, and argumentation as a way to develop deeper understanding of the structure and properties of mathematical concepts and processes (e.g., Franke & Kazemi, 2001; Mulligan, 2011; Mulligan & Mitchelmore, 2009; Scharton, 2004). An important distinction between *instrumental* (procedural) and *relational* (conceptual) understanding (Skemp, 1978) parallels the contrast between reform and traditional approaches to mathematics teaching. Researchers interested in students' conceptual understanding have focused on developing children's awareness of number relationships. Components of number relationships include subitizing, part–whole relationships, and more and less relationships (e.g., Jung, 2011; Jung, Hartman, Smith, & Wallace, 2013). Gray and Tall (1994) distinguish between *procedures* as "things to do" and *concepts* as "things to know" (p. 117), introducing the term *procept* to refer to an amalgam of procedures/processes and concepts (e.g., 6 is the result of counting from 1 to 6 and the result of adding 3 + 3; it is also two groups of three, twice as many as three, and half as many as 12). They argue that more capable children tend to use proceptual thinking (a flexible way of thinking about numbers), whereas less able children persist in using counting procedures.

14.3 Teaching Multiplication and Division

It has been suggested that many parents and teachers do not understand the meaning of multiplication and division and believe that the recall of so-called time tables constitutes knowledge of multiplication (Smith & Smith, 2006). O'Brien and Casey's (1983a, b) findings show that many children in Grades 4 to 6 (9- to 11-year-olds) "do not know what multiplication is" (1983a, p. 250). They are able to do multiplication and division calculations using algorithmic skill, but do not understand what either operation means.

A crucial aspect of work with multiplication is coming to understand the meaning of multiplication as multiple "groups of." At the simplest level, multiplication and division involve three values: the *number* of equal groups, the *size* of these groups, and the overall *total.* If the first two values are known and the *total* is not, the process is multiplication (as in: $2 \times 5 = ?$ meaning two groups of five equal ten). If what is known is the *total* and the *size* of the group and the unknown is the *number* of groups, the process is quotitive or measurement division ($10 \div 5 = ?$). When the *total* and the *number* of groups is known but not the *size* of the groups, the process is partitive or fair-sharing division, the form of division that is most familiar ($10 \div 2 = ?$) to young children and their teachers (Roche & Clarke, 2009).

Evidence shows clearly that children prior to school age can work with equal-group multiplication and fair-sharing division (e.g., Blote, Lieffering, & Ouwehand, 2006; Matalliotaki, 2012; Park & Nunes, 2001; Squire & Bryant, 2003). Hence, it makes sense to capitalize on that prior knowledge in junior mathematics classrooms. The introduction of multiplication and division word problems provides the opportunity to work with units other than one (equal groups), but still leaves open the possibility that children can solve these problems using their preferred strategy, whether it be counting all, counting on, repeated addition, repeated subtraction, or some form of multiplication or division. Even those using less sophisticated strategies may still learn something important about units greater than one. Our study set out to explore the impact on students' mathematics learning of using multiplication and division contexts familiar to children in their early years of schooling.

14.4 Classroom-Based Research

A two-year design research project explored the impact of using multiplication and division problem-solving contexts with five- to seven-year-olds on their emerging understanding of number, including number-fact knowledge, part–whole relationships, and problem-solving strategies. This study was set in an urban school (medium socioeconomic status [SES]) in New Zealand. The participants were 84 five- to seven-year-olds (42 girls & 42 boys) in four classes, two designated as Year 1, one Year 2, and one Year 3. The average age of the students at the beginning of the study was 6.3 years (range 5.0–7.9 years). The children were from a diverse range of ethnic backgrounds, with approximately one-third of European ancestry, one-third Māori (the indigenous people of New Zealand), one-fifth Asian, and the rest were Pasifika and African. One-fifth of the children had been identified as English Language Learners [ELL]. At the start of the study, the children were assessed individually using a diagnostic task-based interview designed to explore their number knowledge and problem-solving strategies (April) (Young-Loveridge & Bicknell, 2015). The assessment interview was completed again after two four-week teaching blocks (September). The assessment tasks included addition, subtraction, multiplication, and division problems; recall of number facts; incrementing in tens; counting sequences; and place value.

14.4.1 Teaching Using Multiplication/Division Contexts

Two series of 12 focused lessons were taught by teachers and researchers; the first phase was in May and the second in August. In these lessons, the children were introduced to groups of two, five, and ten in the context of multiplication, partitive division, and quotitive division. We introduced the children to groups of *two*, using familiar contexts such as pairs of socks, shoes, gum boots, and mittens. Multiplication and division were introduced using simple word problems, such as:

Three children each have a pair of shoes. How many shoes do the children have altogether?

Pene has 12 socks. He puts the socks into pairs. How many pairs of socks are there?

We also used odd-numbered quantities as part of quotitive division into groups of two, in order to introduce the children to the idea of "the odd one out" and give them experience with remainders (Carpenter, Fennema, Franke, Levi, & Empson, 2015). This provided a concrete illustration of odd numbers that made sense to the children.

Once children were familiar in working with groups of two, groups of *five* were introduced, using contexts such as gloves, focusing on the number of fingers on each glove, and five candles on a cake. Groups of five are particularly salient because of their connection to the fingers on each hand. It is also a group size that many children can subitize (quantify without counting one-by-one). In working with groups of five, we also drew children's attention to the way that two groups of five together make ten.

There are 4 cakes on the table. Each cake has 5 candles. How many candles are there altogether?

The lessons then focused on problems using groups of *ten* and the context of filling cartons with eggs or trays with chocolates (each carton or tray had 10 compartments).

We have 20 eggs. We put 10 eggs in each carton. How many full cartons do we have?

A typical lesson began with all children completing a problem together on the mat, using materials to support the modeling process, and sharing ways of finding a solution to the problem. The teacher recorded children's problem-solving processes (including use of manipulatives) and their mathematical ideas in a large scrapbook ("modeling book"). The problem for the day was already written in the book, and both drawings and number sentences were recorded, acknowledging individual children's contributions. The children then completed a problem in their own project books, choosing a similar or larger number, and/or selecting a new number. Materials were made available, and children were encouraged to show their thinking using representations and to record matching equations.

The children improved on all of the assessment tasks given in individual interviews. These included word problems involving addition, subtraction, multiplication, partitive division, and quotitive division. Children's knowledge of number sequence, recall of known facts, and place-value understanding also improved. Here, we focus on the results of the five-year-olds in their first year of formal schooling: the number of five-year-olds who could solve 3 + 4 (objects were shown briefly then screened) more than doubled (from 37 to 80%). By the end of the study, the majority of five-year-olds used either a "count all" (37%) or "count on" strategies (31%). Likewise, performance on 8 + 5 (beans screened under pieces of cardboard) also increased substantially (from 6 to 31%), and most of these children (23% of the total) used counting on. These findings support our conjecture that the focus on multiplication and division provided many opportunities to strengthen children's ideas about addition and subtraction as they were asked to check their answers using a variety of operations, including repeated addition and repeated subtraction.

There was notable improvement in five-year-old children's performance on multiplication involving groups of two (six flax baskets each with two shells inside: from 23 to 86%). Skip counting by two was the most popular strategy (46%), followed by "count all" (36%). Five-year-olds also improved on multiplication with groups of five (a picture of four monkeys each with five bananas: 23 to 63%). Again, skip counting by fives was the most popular strategy. Multiplication with groups of ten improved similarly (three rows of ten cupcakes: from 31 to 66%), with skip counting by tens the most popular strategy.

Quotitive division into groups of five improved substantially (from 9 to 63%) for these five-year-olds. Putting ten socks into pairs was slightly more difficult, but almost half of the five-year-olds could do this by the end of the project. Halving a quantity also improved, with performance varying according to the size of the quantity: half of four, followed by half of eight (with objects: 43 to 66%), 20 (without objects: 9 to 34%), and 100. Partitive division of 8 shared among 4 went from 40 to 66%.

14.4.2 Number Knowledge

Children's knowledge of number-word sequence also improved substantially. For example, the percentage of five-year-olds who could recite the number-word sequence to at least 30 increased from 63 to 91. Those who could recite the sequence to 100 increased from 31 to 74. Skip counting in multiples also improved. For example, initially only 9% of five-year-olds could skip count by twos to 20, but by the end of the study this had increased to more than half. Skip counting in fives to 50 was similar, with improvement from 3 to 51%. Skip counting by tens seemed somewhat easier, with 17% able to count to 100 initially, increasing to almost two-thirds.

In addition to reciting complete number sequences, the five-year-olds became better at identifying the number that is "one more than" or "one less than" a particular number. More than half of them could initially give the number "one more than" a single-digit number, and by the end of the project, this had increased to at least 80%. With two-digit numbers (e.g., 19, 29, 99), there was similar improvement. Five-year-olds improved substantially in their recall of doubles facts up to 10 + 10. Recall of facts involving the addition of one also improved.

14.4.2.1 Place Value

A variety of tasks involving quotitive division into groups of ten were used to assess emerging understanding of place value, for example, answering a question about the number of $10 notes needed to buy a toy costing $80; saying the number of bundles of ten sticks that could be made from a bagful of 60 sticks; identifying the number of cartons of ten eggs that would be full when 23 eggs were put into cartons of ten; connecting the "2" in "24" with two sticks composed of ten Unifix blocks (5 each of contrasting colors); locating the 31st bead in a string composed of groups of ten in alternating colors; and using $10 notes to create a pile of $31 as quickly as possible. Recall of number facts related to place-value understanding also improved (e.g., 20 + 7: from 6 to 46%; 10 + 8: 3 to 46%). Subitizing ten-frames involving groups of ten also improved (e.g., 2 ten-frames: from 14 to 77%; 3 ten-frames: from 0 to 54%).

14.5 Making Connections

There were many connections that occurred as part of teaching multiplication and division with these young children. The first type of connection made was between informal experiences at early childhood or at home involving naturally occurring pairs or "groups of" and more formal experiences of multiplication and division at school. These naturally occurring pairs were explored in relation to body parts such as "a pair of eyes," two arms, legs, ears, and pairs of objects such as mittens, shoes, and socks.

Everyday routines in the classroom outside of the mathematics lesson also provided informal opportunities to link to the mathematics of multiplication. One of the teachers used the context of taking the roll in the morning to explore relationships among numbers. After marking off her roll, she asked the children to name the children who were absent on the day and then to work out how many that was in total. She then asked them to work out how many children were present on that day by subtracting the number absent from the total number of children in the class. To check that calculation, she then counted the children in twos, physically touching two children at a time as she counted up the total. To help the children understand the meaning of an "odd number" (when the total was odd), after she had counted the last pair of children, she asked the class "Are there *two* [Johns]? No, there is only one [John]" and demonstrated how the counting sequence needed to increase by only one when there was an odd number of children. She connected this sequence knowledge with the numerals signifying quantity by pointing to a structured number line displayed on the wall of the classroom. This is consistent with calls by other writers for teachers to "embed learning experiences in the daily routine that engage young children in real-life activities" (Linder, Powers-Costello, & Stegelin, 2011, p. 30).

Another connection was between ordinality (ordering, sequencing) and cardinality (determining how many). The teachers emphasized the nature of number relationships such as "one more than" and "one less than" rather than focusing on the number "just after" or "just before" a particular number. The reason for this particular emphasis was to support children's conceptual understanding of number relationships (Jung, 2011; Jung, Hartman, Smith, & Wallace, 2013), rather than simply focusing on the counting sequence. Initially, children need to learn the sequence of number words assigned to objects during enumeration, but it takes some time before they appreciate that the last number word in the sequence tells how many objects altogether (the cardinality principle: Gelman & Gallistel, 1978).

The connections between composing and decomposing were made to support the development of part–whole thinking. For example, 4 groups of 5 can be recomposed as 2 groups of 10 by children who understand how to use $5 + 5 = 10$ to join 2 groups of 5. Two groups of 10 can be composed additively as a doubles problem ($10 + 10$) as well as multiplicatively as 2 groups of 10 ($2 \times 10 = 20$). The connection can then be made when 20 is decomposed using place-value structure, into 2 groups of 10.

Using multiplication and division contexts that were familiar to the children led to the use of many different strategies for solving word problems, highlighting many mathematical connections among operations and processes. These included the connection of multiplication to repeated addition, skip counting, and counting one-by-one. Using division problems led to making explicit connection of division to repeated subtraction and repeated addition, as well as to skip counting and counting one-by-one. In checking solutions for division problems, connections were also made between division and multiplication (inverse operations).

The children were encouraged to make connections between the modeling process as they worked out a solution and recordings that reflected their actions. The teacher initially recorded these actions using pictures, words, and various equations, and then, the children worked independently on follow-up problems replicating the process. The children's drawings varied considerably from carefully drawn objects (including patterns on socks) to more abstract drawings where a circular shape or tally mark represented the objects in the problems. When solving problems with groups of ten, the ten-frame was drawn to show, for example, 10 eggs in a carton. Children were then encouraged to support these representations with mathematics equations so that they were writing "like a mathematician" using symbols. The connections among the representations provided valuable insight to children's thinking about the quantities.

14.6 Conclusion

We have shown how using familiar contexts with young children and multiplication and division problems contributes to valuable development in mathematical thinking. In some early childhood education (preschool) curricula, there has been a conscious decision to not specify any mathematical content, per se, but instead to emphasize the enjoyment children can experience by "playing around" with quantity, space, and number (e.g., Ministry of Education, 1996). This may reassure teachers that they are addressing the learning needs of young children and avoid the message that early childhood teachers should sit children down for formal didactic mathematics lessons. However, a lack of emphasis on particular mathematical concepts means there are limited opportunities to address teacher beliefs, attitudes, and confidence toward mathematics that has been documented among early childhood education teachers and elementary teachers in the early years (e.g., Chen, McCray, Adams, & Leow, 2014; Gresham, 2007; Vinson, 2001; Wilkins, 2008). Teachers could be further supported through professional learning and development, to become more aware of the ways that they could draw children's attention to key mathematical ideas, referred to by Askew (2013) and Hurst and Hurrell (2014) as the "big ideas" of number.

We need to further examine the view that mathematical expectations for our young children may be too demanding (e.g., Main, 2012). We acknowledge this concern, but our research has shown that if teachers make connections to young children's experiences with "groups of," which is essentially multiplication and division, then their understanding of number is enhanced. The authentic problem-solving situations encouraged children to work with larger quantities than is the norm and evidence showed improvement in number knowledge, addition and subtraction, early place-value understanding, as well as multiplication and division.

References

Askew, M. (2013). Big ideas in primary mathematics: Issues and directions. *Perspectives in Education, 31*(3), 5–18.

Australian Curriculum, Assessment and Reporting Authority [ACARA] (2011). *The Australian curriculum: Mathematics*. Retrieved from: http://www.australiancurriculum.edu.au/Mathematics/Curriculum/F-10.

Bakker, M., van den Heuvel-Panhuizen, M., & Robitzsch, A. (2014). First-graders' knowledge of multiplicative reasoning before formal instruction in this domain. *Contemporary Educational Psychology, 39,* 59–73.

Baroody, A. (2011). Learning: A framework. In F. Fennell (Ed.), *Achieving fluency: Special education and mathematics*. National Council of Teachers of Mathematics: Reston, VA.

Baroody, A. J., Bajwa, N. P., & Eiland, M. (2009). Why can't Johnny remember the basic facts? *Developmental Disabilities Research Reviews, 15,* 69–79.

Behr, M. J., Harel, G., Post, T., & Lesh, R. (1994). Units of quantity: A conceptual basis common to additive and multiplicative intents. In G. Harel & J. Confrey (Eds.), *The development of multiplicative reasoning in the learning of mathematics* (pp. 121–176). Albany, NY: State University of New York Press.

Blote, A. W., Lieffering, L. M., & Ouwehand, K. (2006). The development of many-to-one counting in 4-year-old children. *Cognitive Development, 21*(3), 332–348.

Buchholz, L. (2004). Learning strategies for addition and subtraction facts: The road to fluency and the license to think. *Teaching Children Mathematics, 10,* 362–367.

Carpenter, T. P., Fenemma, E., Franke, M. L., Levi, L., & Empson, S. B. (2015). *Children's mathematics: Cognitively guided instruction* (2nd ed.). Portsmouth, NH: Heinemann.

Chen, J.-Q., McCray, J., Adams, M., & Leow, C. (2014). A survey study about early childhood teachers' beliefs and confidence about teaching early math. *Early Childhood Education Journal, 42,* 367–377.

Clements, D. H., & Sarama, J. (2007). Effects of a preschool mathematics curriculum: Summative research on the *Building Blocks* project. *Journal for Research in Mathematics Education, 38,* 136–163.

Clements, D. H., & Sarama, J. (2009). *Learning and teaching early math: The learning trajectories approach.* New York, NY: Routledge.

Cobb, P. (2007). Putting philosophy to work: Coping with multiple theoretical perspectives. In F. K. Lester (Ed.), *Second handbook of research on mathematics teaching and learning* (pp. 3–38). Reston, VA: National Council of Teachers of Mathematics.

Common Core State Standards Initiative [CCSSI]. (2010). *Common Core State Standards for Mathematics.* Washington, DC: The National Governors Association Centre for Best Practices and the Council of Chief State School Officers. Retrieved from: http://www.coreStandards.org.

Commonwealth of Australia. (2009). *Belonging, being and becoming: The early years learning framework for Australia.* Canberra, ACT: Author. Retrieved from: https://www.coag.gov.au/sites/default/files/early_years_learning_framework.pdf.

Confrey, J., & Harel, G. (1994). Introduction. In G. Harel & J. Confrey (Eds.), *The development of multiplicative reasoning in the learning of mathematics* (pp. vii–xxviii). Albany, NY: State University of New York Press.

Davis, G. E., & Pitkethly, A. (1990). Cognitive aspects of sharing. *Journal for Research in Mathematics Education, 21,* 145–153.

Department for Education. (2013). *The national curriculum in England: Framework document for consultation.* Retrieved from: https://media.education.gov.uk/assets/files/pdf/n/national%20curriculum%20consultation%20-%20framework%20document.pdf.

Department of Education. (2014). *Statutory framework for the early years foundation stage: Setting the standards for learning, development and care for children from birth to five.* Retrieved from: https://www.gov.uk/government/uploads/system/uploads/attachment_data/file/335504/EYFS_framework_from_1_September_2014__with_clarification_note.pdf.

Fischer, F. E. (1990). A part-part-whole curriculum for teaching number in the kindergarten. *Journal for Research in Mathematics Education, 21,* 207–215.

Franke, M. L., & Kazemi, E. (2001). Learning to teach mathematics: Focus on student thinking. *Theory into Practice, 40,* 102–109.

Frydman, O., & Bryant, P. (1988). Sharing and the understanding of number equivalence by young children. *Cognitive Development, 3,* 323–339.

Gelman, R., & Gallistel, C. (1978). *The child's understanding of number.* New York, NY: Cambridge University Press.

Gray, E. M. (1991). An analysis of diverging approaches to simple arithmetic: Preference and its consequences. *Educational Studies in Mathematics, 22,* 551–574.

Gray, E. M., & Tall, D. O. (1994). Duality, ambiguity, and flexibility: A "proceptual" view of simple arithmetic. *Journal for Research in Mathematics Education, 25,* 116–140.

Gresham, G. (2007). A study of mathematics anxiety in pre-service teachers. *Early Childhood Education Journal, 35,* 181–188.

Hattie, J., & Yates, G. C. R. (2014). *Visible learning and the science of how we learn.* London, UK: Routledge.

Henry, V. J., & Brown, R. S. (2008). First-grade basic facts: An investigation into teaching an accelerated, high-demand memorization standard. *Journal for Research in Mathematics Education, 39,* 153–183.

Hopkins, S. L., & Lawson, M. J. (2002). Explaining the acquisition of a complex skill: Methodological and theoretical considerations uncovered in the study of simple addition and the moving-on process. *Educational Psychology Review, 14,* 121–154.

Hurst, C., & Hurrell, D. (2014). Developing the big ideas of number. *International Journal of Educational Studies in Mathematics, 1*(2), 1–18.

Jung, M. (2011). Number relationships in preschool. *Teaching Children Mathematics, 11,* 550–557.

Jung, M., Hartman, P., Smith, T., & Wallace, S. (2013). The effectiveness of teaching number relationships in preschool. *International Journal of Instruction, 6,* 165–178.

Lamon, S. (1994). Ratio and proportion: Cognitive foundations in unitizing and norming. In G. Harel & J. Confrey (Eds.), *The development of multiplicative counting in the learning of mathematics* (pp. 89–120). New York, NY: State University of New York Press.

Lamon, S. (1996). The development of unitizing: Its role in children's partitioning strategies. *Journal for Research in Mathematics Education, 27,* 170–193.

Linder, S. M., Powers-Costello, B., & Stegelin, D. A. (2011). Mathematics in early childhood: Research-based rationale and practical strategies. *Early Childhood Education Journal, 39,* 29–37.

Main, L. F. (2012). Is it too much too soon? Common Core Math Standards in the early years. *Early Childhood Education Journal, 40,* 73–77.

Matalliotaki, E. (2012). Resolution of division problems by young children: What are children capable of and under what conditions? *European Early Childhood Education Research Journal, 20,* 293–299.

Ministry of Education. (1996). *Te Whāriki: He Whāriki Mātauranga mō ngā mokopuna o Aotearoa: Early childhood curriculum.* Wellington, NZ: Author.

Ministry of Education. (2007). *The New Zealand curriculum.* Wellington, NZ: Author. Retrievable from: http://nzcurriculum.tki.org.nz/Curriculum-documents/The-New-Zealand-Curriculum.

Ministry of Education. (2009). *New Zealand curriculum mathematics standards.* Wellington, NZ: Author. http://nzcurriculum.tki.org.nz/National-Standards/Mathematics-standards.

Mulligan, J. (2011). Towards understanding the origins of children's difficulties in mathematics learning. *Australian Journal of Learning Disabilities, 16*(1), 19–39.

Mulligan, J. T., & Mitchelmore, M. (2009). Awareness of pattern and structure in early mathematical development. *Mathematics Education Research Journal, 21*(2), 33–49.

National Council for Curriculum and Assessment [NCCA]. (1999). *Primary school curriculum: Mathematics.* Retrieved from: http://www.curriculumonline.ie/en/Primary_School_Curriculum/Mathematics/Mathematics_Curriculum_.pdf.

National Association for the Education of the Young Child. (2009). *Where we stand on early childhood mathematics: NAEYC & NCTM.* Retrieved from: http://www.naeyc.org/files/naeyc/file/positions/ecmath.pdf.

National Association for the Education of the Young Child [NAEYC]. (2010). *Early childhood mathematics: Promoting good beginnings.* https://www.naeyc.org/files/naeyc/file/positions/psmath.pdf.

O'Brien, T. C., & Casey, S. A. (1983a). Children learning multiplication: Part 1. *School Science and Mathematics, 83,* 246–251.

O'Brien, T. C., & Casey, S. A. (1983b). Children learning multiplication: Part 2. *School Science and Mathematics, 83,* 407–412.

Park, J.-H., & Nunes, T. (2001). The development of the concept of multiplication. *Cognitive Development, 16,* 763–773.

Perry, B., & Dockett, S. (2008). Young children's access to powerful mathematical ideas. In L. D. English (Ed.), *Handbook of international research in mathematics education* (pp. 75–108). New York, NY: Routledge.

Roche, A., & Clarke, D. (2009). Making sense of partitive and quotitive division: A snapshot of teachers' pedagogical content knowledge. In R. Hunter, B. Bicknell & T. Burgess (Eds.), *Crossing divides* (*Proceedings of the 32nd annual conference of the Mathematics Education Research Group of Australasia*) (pp. 467–474). Palmerston North, New Zealand: MERGA.

Sarama, J., & Clements, D. H. (2009). *Early childhood mathematics education research: Learning trajectories for young children.* New York, NY: Routledge.

Scharton. (2004). "I did it my way": Providing opportunities for students to create, explain, and analyze computation procedures. *Teaching Children Mathematics, 10,* 278–283.

Skemp, R. R. (1978). Relational understanding and instrumental understanding. *Arithmetic Teacher, 26,* 9–15.

Smith, S. Z., & Smith, M. E. (2006). Assessing elementary understanding of multiplication concepts. *School Science & Mathematics, 106,* 140–149.

Sophian, C. (2007). *The origins of mathematical knowledge in childhood.* New York, NY: Erlbaum.
Squire, S., & Bryant, P. (2003). Children's models of division. *Cognitive Development, 18,* 355–376.
Steinberg, R. M. (1985). Instruction on derived facts strategies in addition and subtraction. *Journal for Research in Mathematics Education, 16,* 337–355.
Vinson, B. M. (2001). A comparison of pre-service teachers' mathematics anxiety before and after a methods class emphasizing manipulatives. *Early Childhood Education Journal, 29,* 89–94.
von Glasersfeld, E. (1995). A constructivist approach to teaching. In L. P. Steffe & J. Gale (Eds.), *Constructivism in education* (pp. 3–15). Hillsdale, NJ: Erlbaum.
Wilkins, J. L. M. (2008). The relationships among elementary teachers' content knowledge, attitudes, beliefs, and practices. *Journal of Mathematics Teacher Education, 11,* 139–164.
Yackel, E. (2001). Perspectives on arithmetic from classroom-based research in the United States of America. In J. Anghileri (Ed.), *Principles and practices in arithmetic teaching: Innovative approaches for the primary classroom* (pp. 15–31). Buckingham, UK: Open University Press.
Yang, M. T.-L., & Cobb, P. (1995). A cross-cultural investigation into the development of place-value concepts of children in Taiwan and the United States. *Educational Studies in Mathematics, 28,* 1–33.
Young-Loveridge, J. M. (2001). Helping children move beyond counting to part-whole strategies. *Teachers & Curriculum, 5,* 72–78.
Young-Loveridge, J., & Bicknell, B. (2015). Using task-based interviews to assess early understanding of number. In C. Suurtamm (Ed.), *Annual Perspectives in Mathematics Education (APME) 2015: Assessment to enhance teaching and learning* (pp. 67–74). Reston, VA: National Council of Teachers of Mathematics.

Authors Biography

Jenny Young-Loveridge is an Associate Professor at Te Kura Toi Tangata Faculty of Education at the University of Waikato, Hamilton, New Zealand. She has taught at the early primary level, as well as undergraduate and postgraduate mathematics education, research methods, and human development, as part of teacher education and social sciences. Her research focuses on the development of mathematical thinking, in particular the development of number sense and part–whole thinking.

Brenda Bicknell is an Associate Professor at Te Kura Toi Tangata Faculty of Education at the University of Waikato, Hamilton, New Zealand. She has taught across various school sectors and year levels (Years 1–12). Brenda currently teaches undergraduate and postgraduate mathematics teacher education courses. Her areas of interest and research include teacher knowledge, the development of number sense and part–whole thinking, mathematics and language, and the mathematically gifted and talented.

Chapter 15
Slow Maths: A Metaphor of Connectedness for Early Childhood Mathematics

Steve Thornton

Abstract This concluding chapter draws on the work of previous chapters to draw out three dimensions of connection. The first dimension is the connection between mathematical ideas that is characteristic of deep understanding and the distinctive epistemological basis of the discipline of mathematics. The second is the connection between mathematics and culture, in which the joy and delight of playing with and discovering mathematical ideas permeate learning. The third is the deep connection between mathematics and the world, in which mathematics both arises from a consideration of real-world problems and models those problems. Such connections take time and are often undervalued in an era of accountability and easily measured standards. The chapter challenges three dominant metaphors of education —those of education as factory, clinic, or race—and proposes a metaphor of *slow maths*, borrowing from the slow food movement in which connections to people, to culture, to history, to knowledge, to learning, to the world, and to self, are paramount.

Keywords Connections · Slow · Disciplinary norms · History
Culture · Real world · Metaphor

15.1 Introduction

The chapters in this book have described the forging of a broad range of connections related to early childhood mathematics. They have encompassed connections to the discipline of mathematics itself, connections to the development of conceptual understanding through representations, visualization, experiential learning and carefully constructed learning trajectories, connections between home and school, and connections to children's background. What they all have in common is a concern to develop deep understanding of important mathematical concepts,

---

S. Thornton (✉)
Australian Academy of Science, Canberra, Australia
e-mail: Steve.Thornton@science.org.au

V. Kinnear et al. (eds.), *Forging Connections in Early Mathematics Teaching and Learning*, Early Mathematics Learning and Development,
https://doi.org/10.1007/978-981-10-7153-9_15

understanding that not only lays the foundations for further learning but that is also significant in the world of the child. The other common message is that the experiences that develop such understanding need to be carefully planned, to have mathematical integrity and to be developed over time. In this chapter, I use the metaphor of slow food, a way of connecting culture and food that is in direct contrast to fast food, to argue for mathematics in early childhood and school settings that is intimately connected to other mathematical ideas, to the culture in which mathematics develops, and to the world in which mathematics is practised.

15.2 On Connections to Other Mathematical Ideas

In their opening chapter, Kinnear and Wittmann make a passionate plea for placing mathematics at the core of early childhood education. They question a preoccupation with play in early years mathematics teaching and the efforts to make mathematics seemingly more attractive through the use of characters or other devices that young children may identify with. They also suggest that an emphasis on children's physical involvement can be misguided, marginalizing the mathematics and reducing mathematical authenticity and integrity. What is central to mathematics, they argue, is its pattern and structure, and activities that fail to emphasize that structure fail to be mathematically rich. Citing Young (2010), they propose that early childhood educators need to rediscover the disciplinary nature of mathematics and respect the structures and connections that have come to characterize mathematics over the centuries.

The discipline of mathematics is characterized by what Maton (2000) terms a *knowledge* mode of legitimation. Building on Bernstein (1990), Maton discusses the relative strength of *classification*, that is the strength of boundaries between categories or contexts, and *framing*, that is the locus of control within a category, on the epistemic and social relations within a discipline. In the case of mathematics, the epistemic dimension is both strongly classified and strongly framed; that is, it is generally clear what counts and is accepted as legitimate mathematics. The field of mathematics is marked by an accumulation of knowledge that has internal coherence and in which a shared view of knowledge establishes agreed criteria for the validity of that knowledge. On the other hand, mathematics is weakly classified and framed on the social dimension. While cultural and social backgrounds may make it more likely that some members of society will contribute to the growth of mathematical knowledge than others, in the end *who* develops knowledge in mathematics is not important. Indeed, the history of mathematics is riddled with stories of seemingly uneducated or unknown people who have made major mathematical discoveries. Maton contrasts fields such as mathematics with others that have a *knower* mode of legitimation. In such fields, *who* develops knowledge matters more than what that knowledge is, and what counts as legitimate knowledge varies over time and is often marked by radical schisms. He argues that these languages of legitimation are more than mere rhetoric; rather, they "represent the basis for competing claims to limited status and material resources" (Maton, 2000, p. 149).

While Maton's (2000) approach might seem like an esoteric argument about the status of mathematics, I suggest that understanding and respecting the distinctive epistemic norms of the discipline is crucial in our work as mathematics educators. One central purpose of mathematics, from the earliest years on, must be to empower students to move toward the generalization and integration of knowledge that is typical of a knowledge structure with a strong internal grammar. This happens when we provide opportunities for young children to develop deep conceptual understanding, when we connect multiple representations to concepts, when we develop geometric reasoning through visualization, when we construct coherent learning trajectories for understanding our base-ten system, and when we help children to mathematize their experiences to bring out core mathematical ideas. The mathematical education of young children is a serious business, one that requires careful attention to disciplinary norms and the forging of deep connections within the discipline.

In their chapter in this book, Bobis and Way describe how Emma, rather than physically covering her desk with books, drew them with pencil and paper arranged as five rows of 12. In her conversation with the teacher, Emma verbalized her reasoning that there were 60 books, because she could "count by fives." She said that she could also "count by twelves," but she didn't know how to do that and fives was easier. Bobis and Way discuss this as "Property Noticing" in their re-analysis of the episode using the Pirie-Kieran dynamical theory for the growth of mathematical understanding. What is striking about the episode is the connections that Emma has already made between the structure of an array, counting by fives, multiplicative thinking, and commutativity. The outer four layers of the Pirie-Kieran theory—Formalizing, Observing, Structuring and Inventizing—can be considered to be primarily about progressively building theories and making mathematical connections that ultimately lead to "new questions which might grow into a totally new concept" (Pirie & Kieran, 1994, p. 67). The theory illustrates the centrality of making mathematical connections in the growth of mathematical understanding. In his classic work, *A Mathematician's Apology*, G.H. Hardy writes:

> The "seriousness" of a mathematical theorem lies, not in its practical consequences, which are usually negligible, but in the significance of the mathematical ideas which it connects. We may say, roughly, that a mathematical idea is "significant" if it can be connected, in a natural and illuminating way, with a large complex of other mathematical ideas. Thus a serious mathematical theorem, a theorem which connects significant ideas, is likely to lead to important advances in mathematics itself and even in other sciences. (Hardy, 1940, p. 16)

Making and growing mathematical connections such as Emma's and developing an appreciation of the connectedness of mathematics necessarily needs time. It is slow work, requiring immersion into deep and sometimes difficult mathematics rather than superficial learning of a set of skills and techniques. It is not that these skills and techniques are not important; however, their ultimate purpose is to give students the tools with which to see structure and pattern and to reason mathematically.

15.3 On Connections to the History and Culture of Mathematics

There is something austere, perhaps even intimidating sounding, about coming to appreciate the rigor and logic of mathematics, or the web of connections that characterize it as a discipline with a knowledge mode of legitimation. Indeed, the mathematician and philosopher, Bertrand Russell, once wrote "[m]athematics, rightly viewed, possesses not only truth, but supreme beauty—a beauty cold and austere" (Russell, 1959, p. 60). Yet there is also an inherent joy and delight in mathematics that can be experienced by young children and seasoned mathematicians alike.

In his speech to young researchers in mathematics at Cambridge University, renowned mathematician, Fields medalist and Abel Prize winner, Sir Michael Atiyah, looked back on 60 years of mathematics, describing "what views I have seen from the heights and what challenges lie ahead for the next generation" (Atiyah, 2010). He suggested there were many more mountain ranges to explore in mathematics and stressed the cultural dimension of mathematics by projecting 28 slides, not one of which contained text. Rather, each contained a photograph of a mathematician whom he had met or learned from. In effect, he was saying, if not explicitly, "You are part of a long history and culture. Your work will add to the rich tapestry that constitutes the field of mathematics." Atiyah hinted at what motivates mathematicians: understanding, curiosity, exploration, and ideas. In some cases, it is the solution of an elusive problem; in others, it is making a contribution to society by solving a socially significant problem. Yet, in almost all cases, it appears that the overwhelming motivation for mathematicians to do mathematics is to add to knowledge; their motivation is intellectual curiosity and a desire to know the truth.

Fostering intellectual curiosity is as much a part of early childhood mathematics as it is a part of academic mathematics. Experiencing the joy and delight of solving a problem, or coming to appreciate the beauty embodied in mathematical pattern or the gifts of Froebel (2005), is something that every child should experience. Sadly, however, one might argue that not only do few experience such joy and delight, but for many their early experiences of mathematics serve as a kind of "vaccination" that ensures they will remain immune from experiencing mathematical joy and delight for life. In the reSolve: Mathematics by Inquiry project (Australian Government Department of Education and Training, 2017) intellectual curiosity and a desire to know the truth are described as a "spirit of inquiry." It is not that everything done in the name of mathematics has to be a major investigation of a significant real-world issue, but rather that whatever is done ought to foster "ponderings, what ifs, seems to be that's, and it feels as thoughs" (Burton, 1999, p. 30). For example, one of the resources for Foundation is based on the children's book "One is a Snail, Ten is a Crab" (Sayre & Sayre, 2006) that creates numbers based on the total number of legs in the picture. Hence a snail alone is the number one, a person is two, and a snail and a person is three. The book, perhaps surprisingly goes from ten (a crab) to twenty (two crabs). Hence, the first inquiry question is "What numbers are missing?" The second is "How might you make those numbers?", and the third is "How many different ways can you make them?" This last question relates directly to the mathematics of partitions. The number 4, for example, can be partitioned in exactly five ways: 4, 3 + 1, 2 + 2, 2 + 1 + 1, and 1 + 1 + 1 + 1. By the time we reach the number 20, it can be partitioned in 627 different ways, and for the number 70, the number of partitions is over 4 million.

Partitioning is a concept introduced in early childhood mathematics that connects to more complex problems. In Chapter 16 of his 1748 work "Introduction to Analysis of the Infinite," the mathematician Leonhard Euler (2012: English translation) gave a generating function that produces by iteration the number of partitions for each natural number. In the 300 years since Euler's work, the theory of partitions has been the subject of ongoing research, including by the Indian mathematician Ramanujan, who, with G.H. Hardy, in 1918, published an asymptotic formula for the number of partitions that very closely matches the exact answer but does not rely on an iterative process. Of course, one will hardly discuss iterative generating functions and asymptotic approximations with young children, but one can certainly explain that the very problem they are working on (the number of ways of making a given number of legs) is a serious mathematical problem that has been worked on over hundreds of years and that continues to be the subject of mathematical research today. Children come to see mathematics as a living discipline, involving characters such as Ramanujan whose life has been so richly described in the book and film *The Man Who Knew Infinity* (Kanigel, 1991; Pressman et al., 2016). In short, like the young mathematicians addressed by Sir Michael Atiyah at his retirement, they can begin to see themselves as part of the rich tapestry of mathematics that has grown and developed in all cultures throughout human history. They, too, can experience the joy and delight of mathematical discovery.

15.4 On Connections to the World

A school principal whose name has long since vanished from memory once remarked to me in passing words to the effect that "all we have to do to encourage children to learn mathematics better is to make it real." This is a not uncommon statement, yet at the same time it is one that is profoundly flawed. What we think is real for children may not be real or relevant in their minds—indeed, reality often resides in the imagination. Nevertheless, mathematics does have "an unreasonable effectiveness" (Wigner, 1960) in explaining and modeling real-world events and phenomena. Perhaps paradoxically, it is the capacity of mathematics to abstract ideas outside the context that gives it this effectiveness. So a crucial aspect of early childhood mathematics is forging connections between mathematics and the world, but not in a simplistic way that suggests we simply have to "make maths real," but in a way that honors the integrity of both the context and the mathematics.

The chapter by Oslington, Mulligan, and Van Bergen in this volume is a wonderful example of how experiences with real-world data can forge deep mathematical understandings and model-building. There is an obvious and meaningful connection to the children's world, as they analyze the drawings of children of similar ages to determine and describe common structural features. This leads to the development of models that predict the age of an unknown artist that can then be tested against known facts. A similar process is at the heart of most modern marketing techniques, particularly those that use Facebook or other social media. Researchers at Cambridge's Psychometrics Centre (Kosinski et al., 2013) created statistical models to predict personal details using Facebook Likes alone. The models proved 88% correct in predicting male sexuality, 95% accurate in determining African-Americans from Caucasian-Americans, and 85% accurate in differentiating Republican from Democrat voters. Similar predictive modeling techniques are used to target-specific advertising campaigns for specific audiences, as well as species recognition apps for mobile phones that identify which flower or bird has been photographed.

The world is both a rich source of mathematics and a rich area for the application of mathematics. In his autobiography "I Am A Mathematician," Wiener (1956) describes how his career was stimulated by his capacity to cross the boundaries of mathematics and physics, leading to the development of new fields of knowledge, particularly cybernetics, a term coined by Wiener himself. On the occasion of his award of the National Medal of Science, the citation read by President Johnson stated "…for marvelously versatile contributions, profoundly original, ranging within pure and applied mathematics, and penetrating boldly into the engineering and biological sciences" (Rosenblith & Wiesner, 1966, p. 33). Wiener describes mathematicians who have no real contact with physics as having a "thin view of mathematics" (Wiener, 1956, p. 359).

For the child, the world is the physical and social environment at home, at play, in the community, and at school. Forging deep and rich connections between these environments and the world of mathematics both develops mathematical understanding and promotes an awareness of the importance of mathematics in the world. The challenge is to capitalize on situations that have mathematical integrity and that have the capacity to move beyond the simplistic rhetoric that "if it is real children will be interested."

15.5 On Metaphors of Education

The role of metaphor is largely unexamined in educational discourse. Rarely are metaphors studied as interactive devices, shaping the way that we see the world and how we structure reality (Lakoff & Johnson, 1980). It is equally rare to see the role of metaphor examined in the discourse about education and, specifically, its role in shaping teachers' and, more generally, society's views of the goals of education is seldom examined. Yet metaphors profoundly affect aspects of education such as policy, pedagogy, the role of the teacher and student, the nature and purpose of curriculum, and the nature of the school as an educational institution (Botha, 2009). Arguably the three dominant root metaphors of education since the advent of compulsory schooling have been the metaphors of *education as production*, *education as a cure* (Cook-Sather, 2003), and *education as a journey* (Turner, 1998). Each of these metaphors generates a set of associated metaphors that together ascribe particular roles to teachers and students, purposes to the curriculum and overarching goals for schooling.

The *education as production* metaphor casts schools as "factories," a conception that has dominated much educational discourse since the mid-nineteenth century, and remains prevalent today (Darling-Hammond & Friedlaender, 2008). Within the "school as factory" students are conveyed from one site of production to another: in the primary school setting at the end of each year, and in the secondary school setting often at the end of each 50-min period. Within the school as factory, teachers are workers or managers, students are products, and the curriculum is the common production line along which students are progressed. Efficiency, compliance, and quality control, exercised in the form of standardized tests of achievement, are valued, while diversity, critique, and initiative are marginalized. A culture of performance and development and an accountability agenda replete with the language of school improvement (Flint & Peim, 2012) derive from the quality control processes of a factory.

The *education as a cure* metaphor casts schools as clinics that "cure" not only the ills of children, but through that, the ills of society. Its initial conception was in the very first religious schools that aimed to cure the innately sinful and depraved nature of humanity (Cook-Sather, 2003), particularly among children from the lower classes of colonial society. The "school as clinic" focuses on identifying the individual needs of each child as she moves toward a state of health captured by an idealized image of the educated person. The curriculum becomes a prescription, differentiated for each patient on the basis of testing by the teacher, who is both diagnostician and therapist. Within the education as a cure metaphor educational research is dominated by the need for evidence-based practice and by evaluation of the effectiveness of various interventions as measured by their effect-size (Hattie, 2013).

Every metaphor has the potential to highlight or obscure aspects of the concept it seeks to illuminate. The "education as production" metaphor highlights efficiency, while the "education as a cure" metaphor highlights effectiveness. Efficiency and effectiveness are worthy characteristics of a system where there is a clearly defined process and goal, but they are responsive rather than generative. The metaphors of education as production or cure offer neither space to question philosophical bases of education, nor to ask *why* or *whether* rather than *how*. Furthermore, as Cook-Sather (2003) argues, both obscure the individual subjectivity of the people that matter most in education: the students themselves. While the education as a cure metaphor appears at face value to add a human dimension to the education as production metaphor, "their underlying premises—that students are quantifiable products to be packaged or diseased beings in need of remedy—…disable and control those within their constructs" (Cook-Sather, 2003, p. 947).

The *education as a journey* metaphor is particularly common in early childhood education. The early years are cast as preparation for formal schooling, with much attention paid to the transition from home or child-care to Kindergarten, and from Kindergarten to school. Indeed, it is hard to write about education without using a journey metaphor. Students progress through levels of schooling where they may be assigned to different tracks according to whether they are *ahead* of or *behind* their peers. Indeed, the very terms *curriculum* and *course* have their origins in the Latin verb *currere,*[1] meaning "to run." Like the production and cure metaphors of education, the journey metaphor highlights worthwhile aspects of education, but it offers equally little space to question the value or goals of the journey itself.

Taken to its extreme, the journey metaphor becomes a metaphor of *education as a race*, which I suggest has become dominant in political rhetoric in the Western world. In this metaphor, what matters above all else is where a student, or even an entire education system, is placed relative to others. Being left behind is to be avoided at all costs. In the USA, the Bush administration's *No Child Left Behind* legislation was replaced by the Obama administration's *Race to the Top* agenda. Although one focuses on students and the other on the system, they are flip sides of the same root metaphor of education as a race. Similarly in Australia, former Prime Minister Kevin Rudd's apology to Australia's Indigenous Peoples undertook to "halve the *widening gap* in literacy, numeracy and employment outcomes and opportunities for Indigenous Australians" (Rudd, 2008, p. 170, emphasis added), while former Minister for Education, Employment, and Workplace Relations and subsequently Prime Minister, Julia Gillard, warned that Australia was "in danger of *losing the education race*" (Franklin, 2012, para 1, emphasis added) and introduced reforms that aimed to see Australia "back in the *top five* by 2025" (Tovey & McNeilage, 2012, para 11, emphasis added).

Like the production and cure metaphors for education, the race metaphor effectively silences any discussion of whether or not the race is worth competing in or how winners or losers will be determined. In a race, there is no time to stop and admire the scenery or to take a diversion to somewhere that might be more interesting. There is no opportunity for the runners (students) or their coaches (teachers) to question the course to be run (curriculum) or to determine an alternative destination. In a race metaphor for education, rigor and challenge are easily equated with the early introduction of demanding topics, which both misrepresents the nature of mathematical rigor and directly contradicts convincing evidence spanning almost 100 years showing that the early introduction of formal algorithms is counterproductive to developing conceptual understanding of arithmetic operations (e.g., Benezet, 1935).

[1]Perhaps tellingly the word alphabetically preceding curriculum in the Oxford Dictionary of English (Stevenson, 2015) is the word curricle. A curricle was a horse-drawn vehicle in Edwardian England, used for example by Mr. Darcy in Jane Austen's Pride and Prejudice (1813/2008). Curricles were designed for competition in races, were almost exclusively owned by well-to-do males and were almost always single-person modes of transportation.

15.6 On the Virtues of Slow

I have argued elsewhere that we need an alternative metaphor for education that directly challenges the metaphor of education as a factory, cure, or race, and propose that we liken education to "slow food" preparation (Thornton, 2014). As I discuss below, *slowness* emphasizes connectedness to history, culture and traditions, strives for diversity rather than uniformity and has an ethical dimension, in that qualities such as elegance and simplicity are valued more than mere correctness.

In 1986, a McDonald's restaurant was opened at the Piazza di Spagna in Rome. Journalist Carlo Petrini wondered why, if there was "fast food," there could not also be "slow food," and organized a demonstration in which he and his followers brandished bowls of penne as weapons of protest (Honoré, 2004). This was the start of an international "Slow Food Movement," which has since spawned offshoots such as slow travel, slow living, slow cities, slow books, and slow parenting. Despite its name, the "Slow Movement" is not, first and foremost, a movement against speed itself. Rather, it is a philosophy that rejects the one-size-fits-all approach to life that emphasizes uniformity, predictability, and measurability. A core philosophy of this one-size-fits-all approach is the rejection of product variability. Rather than expecting people who operate their business to reinvent the wheel, the company simply expects them to "make it turn faster" (McDonalds, n.d.). To maintain quality and uniformity, each restaurant in the chain must follow set formulas and specifications for menu items, set methods of operation, inventory control, bookkeeping, accounting and marketing, and set concepts for restaurant design and layout. These philosophies typify fast. "Fast is busy, controlling, aggressive, hurried, analytical, stressed, superficial, impatient, active, quantity-over-quality" (Honoré, 2004, p. 14). Contrast this with the philosophy of Quay, a well-known and highly regarded Sydney restaurant:

> This philosophy and passion for the rare and unusual has taken [the chef] all over Australia sourcing unique and exquisite ingredients…The result is a spectacularly innovative cuisine which celebrates the diversity of nature, and which describes and explores the contrasts and harmonies in textures and flavours one finds. there (Quay Restaurant, n.d.)

Although not advertising itself as a "slow" restaurant, Quay epitomizes a slow philosophy. "Slow is… calm, careful, receptive, still, intuitive, unhurried, patient, reflective, quality-over-quantity. It is about making real and meaningful connections—with people, culture, work, food, everything" (Honoré, 2004, p. 14).To elaborate further upon the concept of slowness it is helpful to look at the key principles that underpin the slow food movement.

> First, it expresses a definite *philosophical position*—that life is about more than rushed meals. Second, it draws upon *tradition and character*—eating well means respecting culinary knowledge and recognizing that eating is a social activity that brings its own benefits. A respect for tradition also *honors complexity*—most sauces have familiar ingredients, but how they are combined and cooked vitally influences the result. And third, slow food is about *moral choices*—it is better to have laws that allow rare varieties of

> cheese to be produced, it is better to take time to judge, to digest, and to reflect upon the nature of 'quiet material pleasure' and how everyone can pursue it. (Holt, 2002, p. 267, emphases in original)

I will add two further principles: that uncertainty is inherent in the process of creation; and that variability is a quality to be treasured rather than feared.

One could substitute the word "education," or indeed "mathematics," for each of the words pertaining to food in the above description, and the paragraph would make almost perfect sense. A slow education, then, has a clearly articulated philosophical basis; it values culture and tradition; it blends established techniques and fresh ideas in an environment where uncertainty is encouraged; it values variability rather than uniformity; and it has an ethical dimension with which to judge what is good and worthwhile. "Slow Maths" is not simply about taking longer to learn the same set of skills and concepts for the same purposes, although there is ample evidence that students do learn traditional content better by learning it more slowly and deeply (e.g., Boaler, 1997; Nathan, 2009). Rather, it is about a fundamentally different approach to, and mindset about, early childhood and school mathematics. It creates a space which foregrounds mathematics as a way of *being* and *acting* in the world.

15.7 Conclusion

In this concluding commentary, I have outlined some of the dimensions of connectedness that are evident in the chapters of this book and shown how these connections are also evident in the wider world of mathematics. These connections are both within and without the discipline itself—mathematics is characterized by a rich conceptual network in which one concept links powerfully to others, it is characterized by the joy and delight of discovery that has been part of the culture of mathematics for centuries, and it is characterized by deep and meaningful connections to the world. The chapters of this book have encapsulated many of these connections and shown how early childhood mathematics education can take on many of the dimensions that are evident in the discipline of mathematics.

But these connections are often undervalued in educational regimes that focus on accountability, on efficiency, or on effectiveness. The dominant metaphors of education as factory, cure, or race tend to focus on a limited set of easily measurable outcomes that have come to define what is important in education. If we, like the authors of the chapters in this book, are to take seriously our role as early childhood mathematics educators, we need to challenge the narrow views of education that permeate the current educational agenda. One way to challenge these views is to reframe education as being about forging connections. *Slowness* is first and foremost about "connections—to people, to culture, to history, to knowledge, to learning, to the world, and to self" (Slow Movement, 2013).

References

Atiyah, M. (2010). *Young researchers in mathematics 2010*. Retrieved 9 May, 2014, from http://sms.cam.ac.uk/media/753316;jsessionid=30C071FEDDCBD0EBBE3014FB9797C443.

Australian Government Department of Education and Training. (2017). reSolve: Maths by Inquiry. Retrieved 4 July, 2017 from www.resolve.edu.au.

Benezet, L. P. (1935). The teaching of arithmetic I: The story of an experiment. *Journal of the National Education Association, 24*(8), 241–244.

Bernstein, B. B. (1990). *Class, codes and control: The structuring of pedagogic discourse* (Vol. 4). London: Routledge.

Boaler, J. (1997). *Experiencing school mathematics: Teaching styles, sex and setting*. Buckingham: Open University Press.

Botha, E. (2009). Why metaphor matters in education. *South African Journal of Education, 29*(4), 431–444.

Burton, L. (1999). Why is intuition so important to mathematicians but missing from mathematics education? *For the Learning of Mathematics, 19*(3), 27–32.

Cook-Sather, A. (2003). Movements of mind: The matrix, metaphors, and re-imagining education. *The Teachers College Record, 105*(6), 946–977.

Darling-Hammond, L., & Friedlaender, D. (2008). Creating excellent and equitable schools. *Educational Leadership, 65*(8), 14.

Euler, L. (1748). *Introduction to analysis of the infinte, book 1*. (J. Blanton, Trans. 2012). New York: Springer Science and Business.

Flint, K. J., & Peim, N. (2012). *Rethinking the education improvement agenda: A critical philosophical approach*. London: Continuum.

Franklin, M. (2012). *We risk losing education race, Julia Gillard warns. The Australian*. Retrieved from: http://www.theaustralian.com/national-affairs/policy/we-risk-losing-education-race-julia-gillard-warns/story-fn59nlz9-1226251791091.

Froebel, F. (2005). *The education of man* (W. H. Hailmann, Trans.). New York: Dover Publications (Original work published 1898).

Hardy, G. H. (1940). *A mathematician's apology*. Cambridge: Cambridge University Press.

Hattie, J. (2013). *Visible learning: A synthesis of over 800 meta-analyses relating to achievement*. New York and London: Routledge.

Holt, M. (2002). It's time to start the slow school movement. *Phi Delta Kappan, 84*(4), 264–271.

Honoré, C. (2004). *In praise of slow: How a worldwide movement is challenging the cult of speed*. London: Orion Books.

Kanigel, R. (1991). *The man who knew infinity*. New York: Charles Scribner's Sons.

Kosinski, M., Stillwell, D., & Graepel, T. (2013). Private traits and attributes are predictable from digital records of human behaviour. *Proceedings of the National Academy of the United States of America, 110*(15), 5802–5805.

Lakoff, G., & Johnson, M. (1980). *Metaphors we live by*. Chicago: University of Chicago Press.

Maton, K. (2000). Languages of legitimation: The structuring significance for intellectual fields of strategic knowledge claims. *British Journal of Sociology of Education, 21*(2), 147–167.

McDonald's. (n.d.). *Becoming a key ingredient in our success: McDonald's Australia franchising overview*. Retrieved from https://McDonalds.com.au/sites/McDonalds.com.au/files/Franchising%20Overview_0.pdf.

Nathan, L. F. (2009). *The hardest questions aren't on the test: Lessons from an innovative urban school*. Boston, MA: Beacon Press.

Pirie, S., & Kieren, T. (1994). Growth in mathematical understanding: How can we characterise it and how can we represent it? *Educational Studies in Mathematics, 26*, 61–86.

Pressman, E. R., Young, J., Thomas, J., & Montgomery, M. (Producers), & Brown, M. (Director). (2016). *The man who knew infinity*. United Kingdom: Pressman Film, Xeitgeist Entertainment Group, & Cayenne Pepper Productions.

Quay Restaurant. (n.d.). *Quay: Food philosophy*. Retrieved from http://www.quay.com.au/page/food-philosophy.html.
Rosenblith, W., & Wiesner, J. (1966). From philosophy to mathematics to biology. *Bulletin of the American Mathematical Society, 72*(1), 33–38.
Rudd, K. (2008). *Apology to Australia's Indigenous Peoples*. Commonwealth of Australia Parliamentary Debates, House of Representatives Official Hansard, Parliament of Australia. 42nd Parliament, First Session, First Period, Wednesday, 13 February 2008, 9 am, 167–173. Retrieved from: http://parlinfo.aph.gov.au/parlInfo/genpdf/chamber/hansardr/2008-02-13/0003/hansard_frag.pdf;fileType=application%2Fpdf.
Russell, B. (1959). *Mysticism and logic and other essays* (11th ed.). London: George Allen & Unwin.
Sayre, A. P., & Sayre, J. (2006). *One is a snail, ten is a crab*. Cambridge, MA: Candlewick Press.
Slow Movement. (2013). *Slow movement*. Retrieved from http://www.slowmovement.com.
Stevenson, A. (Ed.). (2015). *Oxford Dictionary of English*. Oxford: Oxford University Press.
Tovey, J., & McNeilage, A. (2012). The great race to the top of the class. *Sydenye morning herald*. Retrieved from: http://www.smh.com.au/national/education/the-great-race-to-the-top-of-theclass-20121214-2bexo.html.
Thornton, S. (2014). Slow maths: A metaphor of connectedness for school mathematics. *Doctoral Dissertation*. Australian National University. Retrieved from https://openresearch-repository.anu.edu.au/handle/1885/111190.
Turner, J. (1998). Turns of phrase and routes to learning: The journey metaphor in educational culture. *Intercultural Communication Studies, 7*, 23–36.
Wiener, N. (1956). *I am a mathematician: The later life of a prodigy*. Cambridge MA: MIT Press.
Wigner, E. P. (1960). The unreasonable effectiveness of mathematics in the natural sciences. *Communications on Pure and Applied Mathematics, 13*(1), 1–14.
Young, M. (2010). The future of education in a knowledge society: The radical case for a subject-based curriculum. *Journal of the Pacific Circle Consortium for Education, 22*(1), 21–32.

Author Biography

Dr. Steve Thornton is Executive Director of Mathematics by Inquiry, an Australian government funded project conducted by the Australian Academy of Science in collaboration with the Australian Association of Mathematics Teachers. The project produces classroom and professional resources to promote inquiry approaches to school mathematics from Foundation to Year 10. Steve has previously lectured in mathematics teacher education at the University of Oxford, the University of Canberra, and Charles Darwin University. He has wide experience as a teacher of mathematics, as a researcher in mathematics education, as leader of a national professional development program for the Australian Mathematics Trust, and as a consultant and critical friend to numerous mathematics education projects.

Steve is a Life Member of the Australian Association of Mathematics Teachers and has been awarded a B.H. Neumann medal for excellence in Australian mathematics education. He has written numerous journal articles and book chapters and has presented keynote addresses and workshops at Australian and international mathematics education conferences.

MIX
Papier aus verantwortungsvollen Quellen
Paper from responsible sources
FSC® C105338

If you have any concerns about our products,
you can contact us on
ProductSafety@springernature.com

In case Publisher is established outside the EU,
the EU authorized representative is:
Springer Nature Customer Service Center GmbH
Europaplatz 3, 69115 Heidelberg, Germany

Printed by Libri Plureos GmbH
in Hamburg, Germany